Introduction to Satellite Systems

Technology Basics, Market Growth, Systems, and Services

Ben Levitan, Lawrence Harte

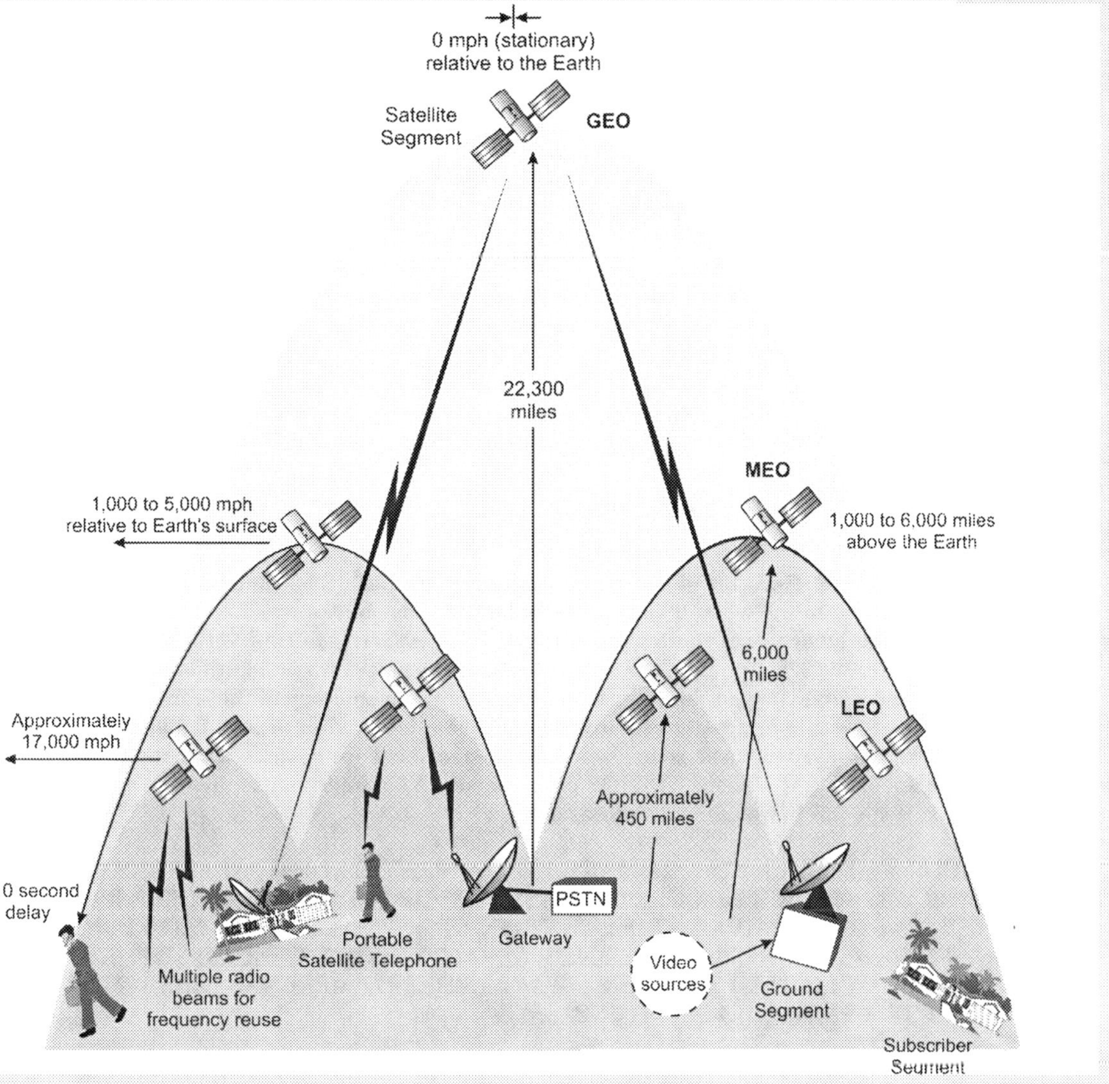

0 mph (stationary) relative to the Earth
Satellite Segment
GEO
22,300 miles
MEO
1,000 to 5,000 mph relative to Earth's surface
1,000 to 6,000 miles above the Earth
6,000 miles
LEO
Approximately 17,000 mph
Approximately 450 miles
1/10 second delay
PSTN
Portable Satellite Telephone
Gateway
Video sources
Multiple radio beams for frequency reuse
Ground Segment
Subscriber Segment

Excerpted From:

Wireless Systems

With Updated Information

ALTHOS Publishing

ALTHOS Publishing

ISBN: 0-9742787-8-5

Some or all of the materials in this ebook also appears in the print version of the book : Wireless Systems ISBN: 0 9728053-4-6

ALTHOS electronic books (ebooks) and images are available for use in educational, promotional materials, training programs, and other uses. For more information about using ALTHOS ebooks and images, please contact Karen Bunn at kbunn@Althos.com or (919) 557-2260

Terms of Use

About the Authors

Mr. Harte is the president of Althos, an expert information provider which researches, trains, and publishes on technology and business industries. He has over 29 years of technology analysis, development, implementation, and business management experience. Mr. Harte has worked for leading companies including Ericsson General Electric, Audiovox/Toshiba and Westinghouse and has consulted for hundreds of other companies. Mr. Harte continually researches, analyzes, and tests new communication technologies, applications, and services. He has authored over 50 books on telecommunications technologies and business systems covering topics such as mobile telephone systems, data communications, voice over data networks, broadband, prepaid services, billing systems, sales, and Internet marketing. Mr. Harte holds many degrees and certificates including an Executive MBA from Wake Forest University (1995) and a BSET from the University of the State of New York, (1990).

Mr. Ben Levitan is an engineer and an expert on new and developing wireless standards for Satellite, Cellular and land mobile radio (LMR) Systems including ANSI-41, GSM, and 3rd Generation systems. Mr. Levitan is an active participant for the US and international development of ANSI-41, GSM, 802.11, All-IP (3G Standards), Interstandard Roaming and FCC and US Government mandated features and requirements. Mr. Levitan has represented COMSAT, Intelsat and Aeronautical Radio, Inc. in the development of satellite standards and was a member of the United States Delegations to the United Nation's International Telecommunications Union conferences for worldwide standardization of telecommunications for eight years. Mr. Levitan has consulted and is an expert witness for several major wireless companies. Mr. Levitan's expertise includes international roaming for cellular systems and government and FCC mandated features such as the new required Wiretap standards (CALEA), Local Number Portability and 911 emergency standards for the cellular industry. Mr. Levitan is a frequent speaker at wireless industry forums and conferences.

Table of Contents

Satellite Systems

Satellite systems provide a unique way to connect communications networks using space vehicle(s) in orbit over the Earth. Satellites simply act as transport for communications much like cables, fiber optics or microwave systems that are the path between two communicators. Similar to cables that provide long distance communications, a satellite acts like a repeater to assure that the signal, video or voice communications that it is transporting remains as near to the original as possible. Repeaters in satellites take the weak signal it receives and restores it before passing back to the recipient.

Satellite systems can provide voice (telephony), dispatch, data, and video broadcast services. Mobile satellite services (MSS) provide for two-way voice and wide area dispatch communication. Very small aperture terminals (VSAT) use satellite to provide high-speed data links almost anywhere in the world. Satellite digital audio radio services (SDARS) are achieving commercial through the use of efficient transmission and innovative pay for subscription audio and multimedia services. Accurate location information can be obtained by global positioning system (GPS). Direct video broadcasting (DVB) satellite systems are projected to provide broadcast television service to more than 90 million subscribers by 2007.

Satellites are unique for communications for many reasons. The start up costs for satellite communications is very high. However, the benefits of using satellite communication can be extremely good. For example, if a satellite is placed in orbit over the earth, it is capable of providing a communication connection between any two points within it's view by simply having a transmitter and receiver at each of the two points (in the form of a satellite dish). The view of a typical satellite in orbit over the earth is 1/3 of the planet. As well, the multiple connections can be established. Depending on the capability of the satellite, thousands of paired connections can be established at one time between two points. The advantage of this over established point-to-point cable connections is a significant benefit of a satellite.

A satellite in space is limited by its design as to how many signals it can receive and retransmit based on the number of "transponders" it holds. However, one of the biggest communications advantages of a satellite is that the number of communication devices on the ground that can receive the signal and use it is not limited. This makes satellites an ideal system for broadcast uses. Even a small satellite with only a few transponders can provide a significant service. Satellite radio systems are simple. A signal is transmitted to a satellite and the satellite, acting as a simple repeater retransmits the signal. If listeners on the earth have a receiver, there is no limit to the number of customers that can receive the signal. For more than 30 years, satellites have been providing voice and data communication service around the globe; however, the cost for equipment and services has been very high.

In 1997, the high cost of satellite equipment and services began to reduce dramatically. New high capacity satellites and digital technology allow for lower cost service and advanced messaging services. Early satellites used analog transmission. After the development of digital satellites, which offer more capacity, several more satellites were put into orbit, followed by the next-generation of low orbiting satellites. These new developments are rapidly bringing the cost of equipment down by over 75%.

Additional factors in reduction of the costs of satellites were the advent of the reusable Space Shuttle. Despite the very public failures of the shuttle, its success rate and cost in delivering satellites to their orbit is dramatically more favorable than the pre-shuttle era. The failure rate for delivery of satellites prior to the shuttle was high. In addition to the shuttle, several countries such as France, Russia, Japan and China are now providing alternative launch "delivery systems" in competition with the shuttle or other USA based delivery systems to get satellites to orbit and this is dramatically reducing costs.

Although not commonly known to those in developed nations, telephone service is not ubiquitous. In fact, the United Nations recognizes only one telecommunication service as now being available in every country in the world. That service is Telex, a slow text transmission service that is nearly forgotten in most developed countries. Since one satellite can cover 1/3 of the earth, three satellites are capable of covering the entire planet and these systems are in place and are providing a way to service these remote areas with telephones, news and information feeds if they have the transmitting and receiving equipment.

Satellites experience very few operational problems once they are placed into orbit in free space where there is little or no air. In such an environment, there are few obstacles that slow the satellites down or wear them out once they are sent into orbit. It is unlikely they will be damaged (such as being dug up by a backhoe) and the equipment is designed to be very reliable. Because they are delivered to satellite via a large vehicle that moves fast and can vibrate violently, satellites are designed to withstand the rough trip into space. Once the satellite reach free space, there is little additional trauma. So, if the satellite is successfully delivered to space and it is operational, there are few reasons to cause equipment failure. Because most satellites are not accessible, even by a space shuttle, once launched into space, they are designed so there are no parts that would wear out and to have no single points of failure. Every system within a satellite has an automatic back up system. While ground based systems can use lower cost parts and assemblies that have a limited lifetime, satellites are not accessible so the parts they use are designed to rarely fail.

Once a satellite is placed in orbit, it could theoretically last forever. However, there are various reasons why a satellite that is placed in a delicate orbit over the Earth has a limited operating lifespan. Satellites tend to drift away from their orbit and turn so that its' transmitter is not pointing towards the Earth. Controlled by signals from monitoring systems on the ground, small rockets on the sides of the satellite can be fired to move it back to it's intended orbital position. As a result the useful lifetime of a satellite is generally more dependent upon fuel reserves than on wear.

Satellites are typically classified by the type and height of the orbit they have been placed in around the earth orbit. A satellite can be placed either in orbit so it appears to be stationary over a part of the earth (geosynchronous) or so it appears to be orbiting the earth (Polar Orbit).

In a geosynchronous orbit, a satellite appears to be stationary over the earth. A satellite in orbit is actually traveling at a high-speed. When we want a satellite to remain over a country to provide communications throughout that country, in reality it must be place in orbit so it is moving at the same speed as the earth. The speed of the earth at the Equator is 1670 km/hr (1035 MPH). A satellite placed in orbit over the equator and traveling around the planet at the equator at 1,035 MPH will appear to be stationary and will be able to provide a communications platform for the earth below it. Remember that once an object in space is in motion with no friction to slow it down it will remain fairly constant, but still, "solar winds", "sun storms" and the fact that the satellite and the earth may be slightly off in speed will cause the satellite to appear to "drift" which needs to be corrected.

In a polar orbit, the satellite appears to pass over a position on earth and move away like an airplane-flying overhead. In a polar orbit a satellite is actually placed in an orbit at a high-rate of speed that passes over one pole of the earth and then over the other. As the earth turns, each time the satellite passes the equator the earth has turned a little and the satellite is passing over a different part of the earth. Each day, the satellite has covered the entire earth but only for a small amount of time. This type of satellite is good for a variety of uses such as photographic missions, spying, or communications in a store and forward mode. As well, if a provider can place a large number of polar satellites over the earth in polar orbits at the same time, and each is in an orbit that is slightly shifted from the next, then the entire planet can be covered at the same time. Motorola's Iridium system attempted to do this. The plan was to have six satellites in the same polar orbits but each one starting and ending their orbit at a different point, so that while one satellite was over the north pole, the next was nearing the equator headed towards the south pole and the next was nearing the south pole moving away from the equator while the fourth, fifth and sixth satellites were on the opposite side of the earth moving towards the north pole. This would allow coverage of an entire band of the earth. By adding 10 more similar bands the entire earth was covered at the same time by 66 moving satellites in polar orbit.

Through the uses of applied physics and trial and error, it was determined that the best altitude for a satellite to be placed in an orbit, that would make it appear stationary over the earth (GEO - Geosynchronous), is approximately at 22,300 miles. This provides the best coverage, the easiest launch and the most stable environment. For satellites in a polar orbit, (Low Earth orbit, or LEO) the ideal orbit is located at approximately 500-1,000 miles above the earth and a single one can cover a thousand miles. Some systems attempt to split the difference and provide a lower cost but higher coverage at a medium earth orbit (MEO). Generally the lower the orbit, the lower the launch costs, but the higher the orbit, the more of the earth that can be viewed. MEO satellites are commonly positioned up to 6,000 miles above the earth and a single one can cover several thousand miles.

Figure 1.1 shows that a geostationary satellite is launched from Earth to achieve an orbit of approximately 22,300 miles above the Earth. At this position, it appears to have the same relative position above the surface of the Earth. This diagram shows that the combined effects of speed and gravity actually keep the satellite in the same relative position above the earth.

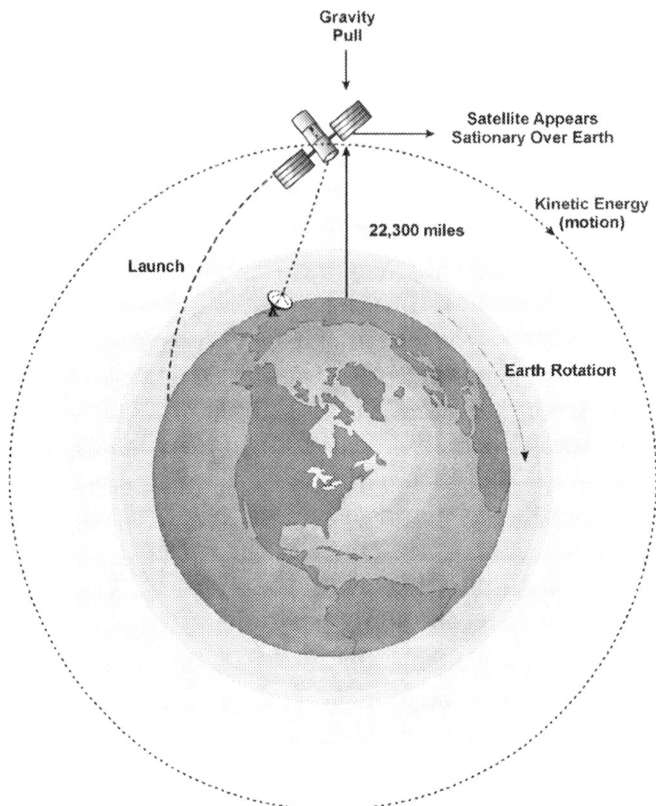

Figure 1.1, Geostationary Satellite Orbit

There are two key parts to a satellite system; The Telemetry, Tracking and Control systems and the Communication Payload. The Telemetry, Tracking and Control refer to the ground station and the portions of the satellite responsible for getting it into orbit and keeping it there. The Communication Payload refers to the transponders or other equipment responsible for serving the customers.

There are three basic types of satellite telecommunications services: broadcast, multicast and trunking. Broadcast and multicast provide service from a single provider on the ground to either "everybody" or various other fixed points. Trucking is point to point as in telephone communications.

The specific frequencies dedicated to satellite services are in the microwave bands (such as the C band, K subscript u (Ku), K subscript a (Ka) and L bands). The FCC and other global entities control these frequencies. Satellite transmissions cover large parts of the earth and can tend "encroach" on another country. A general policy is that every country controls their own "sky" and is their own authority for satellite transmission within their country. A company trying to develop a satellite system to cover multiple countries must have an "in country" operating authority to run the satellite system within the country. Motorola's Iridium system, Intelsat and other "world providers" of satellite systems have hundreds of operating agencies, each homed in the different countries to support such systems.

With so many satellites in space, one might wonder why they don't occasionally "run into each other". The orbit for most geostationary satellites is 22,500 miles. These make for an orbit of 141,000 miles (remember the circumference of a circle is pi times the diameter of the circle.) The odds of a satellite that is the size of a speck of dust in comparison to this circumference running into another one is astronomical. None the less, to prevent any electromagnetic interference or other station keeping problems, satellites are voluntarily spaced at least 2 degrees away from each other in orbit.

The Satellite

A satellite is a space vehicle that orbits the earth, and which contains one or more radio transponders that receive and retransmit signals to and from the earth. A satellite contains both Communications Payload and Telemetry, Tracking and Control systems. The size and weight of satellites varies from 1 meter to over 20 meters in length and 90 lbs (40 kg) to over 8800 lbs (4000 kg). There are several 4000 kg GEO satellites that fill a large portion of the space shuttles payload bay.

The major components of the satellites include:

Power supply system - Batteries and solar panels that provide power for all the systems within the satellite. The amount of power used by satellites varies from a few hundred watts for small low earth orbit satellites to several thousand watts for large high earth orbit satellites. For large satellites with large power needs, nuclear energy is an excellent solution. If the system can be safely delivered to space there is little worry of human harm that could come from a failed nuclear reactor.

Position control system – Systems that send position information from itself to the ground and also responds to commands from the ground regarding position. Commands from the ground will call for short firing of the side jets to move the satellite.

Transmitters and receivers – These systems known as transponders are the repeaters, which receive data from the antenna system and return it to the antenna for transmission, back to the earth.

Antennas – Each satellite has at least two antennas. The Omni is an antenna that broadcasts in any direction. This is generally used for command and control to send commands to the satellite when it is being placed into orbit. No matter what the orientation of the satellite, it will be able to send or receive commands from the earth. The Main antenna is for communications. It is a focused single directional antenna that points to the desired communications area.

Figure 1.2 shows how a basic satellite transponder operates. In this example, the transponder receives a signal from an earth station. This signal is translated (converted) into a higher frequency, amplified, and re-transmitted back to the Earth.

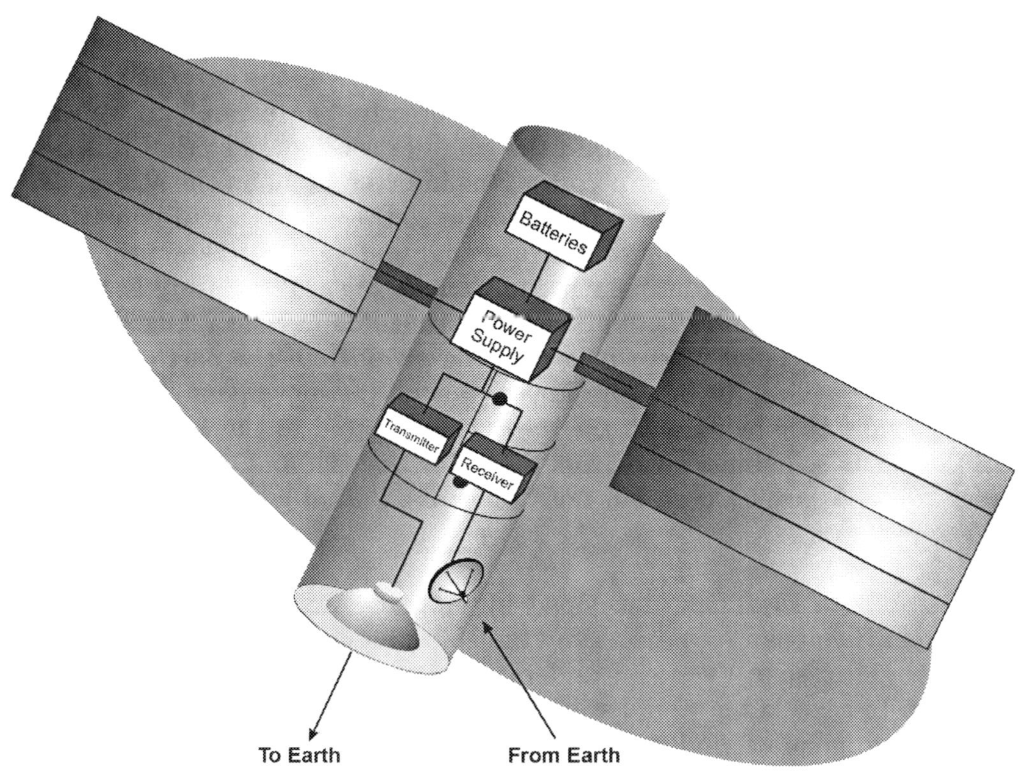

Figure 1.2, Basic Satellite Transponder

There are two general types of satellites; Spin stabilized and Three-Axis satellites.

Three Axis satellites are satellites, which have small rockets that can gently "move" it in one of three directions. Since there is no "up" or "down" and as discussed, the satellite is actually moving at a high rate of speed in orbit, these three rockets actually change the speed of the satellite in three directions. Since the goal of good satellite communications is to provide coverage over a portion of the earth, it's best to think of direction in terms of earth. The three-axis satellite has one rocket (small jet) that speeds the satellite up if it is not traveling fast enough around the equator to stay over the coverage area. There is also a matching jet on the opposite side of the satellite to slow it down if it is getting ahead of its intended coverage area.

Remember in space, if you push something, it moves and won't stop. So a gentle push by the jets will move the orbit faster or slower. Assuming that the satellite is perfectly moving around the earth over the coverage area, it is possible that the antenna is pointed incorrectly. For example, if we want to cover the USA it is possible that the satellite is oriented so that it is over the US but the antenna is pointed to Canada. A small jet on the second axis of the satellite will gently turn the satellite so the antenna will point down and the opposite jet will stop the turn. A slight jet firing will start the satellite's antenna moving from pointing to Canada towards pointing to the US but unless opposite force is applied it will continue to South America then turn over to the point where it is pointing away from the earth and then back around again. The opposite force will stop the movement. Finally, the satellite may be moving with the earth and pointed to the latitude of the USA but may be pointed away from the US to the east and be providing coverage to the Atlantic Ocean but not California. Paired jets on the third axis will correct this.

Spin stabilized satellites, developed by Hughes Aircraft uses a different approach. If you can visualize a toy top spinning perfectly on the sidewalk (as it would in space) you can understand the concept of a spin-stabilized satellite. In space a top would spin perfectly. Using some clever design, the communication platform is on top of the satellite but doesn't spin. By design it points in one direction. If you touched the top of a spinning top in space it would lean a little in the direction of the push. This is the concept. Small jets can push the top to face earth and it stays more stable than a three-axis satellite.

Telemetry, Tracking and Control (TT&C)

Each time a satellite's jets are fired to move a satellite; some of the stored fuel is burned. If the satellites fuel is burned up before it is placed in the correct orbit, the satellite is lost. A satellite must have enough fuel to get it to the right orientation in space and also skilled technicians must carefully fire the rockets to get it there without using all the fuel.

Satellites do not automatically stay in their desired location, because of gravity effects from the irregular earth, sun, moon, and other planets. The position of the satellites must be continuously monitored and adjusted (an activity called "station keeping"). Each station-keeping event burns a little fuel. This means that the life of the satellite system is usually determined by the amount of fuel it can carry. Once there is no fuel left for station keeping the satellite will drift until its antenna is no longer oriented correctly and communications ability is lost. The controllers know this and the life of the satellite can be carefully determined to avoid a loss of service. The satellite can be replaced before it expires. If a satellite is allowed to simply "run out of gas", it will drift aimlessly out of position but occasionally will repoint to the earth. At this time, if it is transmitting it could easily disrupt communications with the customers of the replacement satellite. For this reason, before a satellite runs out of fuel, the controllers usually use the last of the fuel to push the satellite to a position over the Atlantic Ocean where it is unlikely to cause any disruption or accidentally run into another satellite. The systems are turned off as well.

There are a number of strategies and tricks for keeping the satellite going as long as possible. The life of a geosynchronous satellite is typically 15 years and a LEO about five years. COMSAT Labs in the 1980's noticed that the majority of drift for a spin-stabilized satellite tends to be in the "North-South" direction in relation to earth rather than "East-West". Satellite controllers were doing 100 times more than station keeping for North-South corrections. COMSAT patented a strategy, which takes advantage of this knowledge. Allowing the satellite to drift in the North-South direction and programming the ground antennas to simply track the satellite in the North-South direction as it drifts predictable North-South can save fuel saved by not constantly correcting the drift or "wobble". Relative to earth, the satellite would appear to wobble (hence the term "WobbleSat") and the antennas simply track the movement to maintain communications. The satellite is corrected at a less frequent rate and the fuel saving is substantial since the correction is only performed when the drift is getting out of the tracking range of the antennas.

Most satellites in existence today are three-axis stabilized. The ability to control the attitude stability (rotation) of these satellites is normally controlled by momentum wheels, magnetic torques, and sometimes-small rocket motors for three-axis stabilized satellites. Spin-stabilized satellites, of which there are relatively few in the GEO type, are stabilized by the gyroscope effect of spinning. Electric motors and earth sensors stabilize the payloads.

The main purpose of satellites is to receive radio signals from earth and retransmit these signals back to earth. Usually, this is accomplished by a transponder. Such a transponder is called "bent-pipe." The transponder receives a signal on one frequency, converts the frequency, amplifies the signal and then sends the signal back to earth. There may be 40 or more transponders on a single satellite, each having a radio channel bandwidth of 80 MHz (or more) some modern satellites today do on-board digital processing of the signals. These satellites are much more complex and expensive than the simple transponder types described above.

The antenna system of a satellite usually consists of several directional antennas, allowing the satellite to direct its radio energy to specific locations. The radio coverage area provided by a satellite is called its footprint. The footprint of a single satellite can be thousands of miles in diameter, which may be enough to cover an entire continent. Sometimes, however, spot beams are used that cover only a hundred miles or so in order to concentrate signals into a small area and develop stronger signals because of higher antenna gain.

In general a satellite service consists of all the satellites an operator has in their inventory. LEO systems may have as many as 66 satellites while a Direct TV system may operate only a single satellite. As well, a satellite may serve a single customer, such as an XM radio system operated by Boeing for XM radio, or multiple customers such as Intelsat who serves TV broadcasters, long distance telephone and private satellite links all from one satellite.

Ground Segment

The ground segment of a communication satellite system contains the gateways that send and receive information signals from the satellite, switching or routing facilities and a satellite control center. Again, there is TTC equipment and Communication Equipment.

The communication equipment consists of gateways that provide access to the space segment and interface to public and private data networks. The major elements of a gateway include gateway earth stations, each of which is composed of an antenna, controller and radio equipment. The systems used in a gateway typically have redundant assemblies to allow rapid restoration of service in the event of failure. A satellite system always has two command and control stations and gateways. For MEO and LEO systems, gateways are located throughout the world. These facilities monitor and manage all network elements to ensure continuous, consistent operations in the provision of quality service.

Gateways connect with a media source or switching system that provides interconnection to the terrestrial networks. A media source may consist of television channels that come from a nearby studio or via network information feed. Switching systems for mobile satellite systems (MSS) are used to connect and process communication paths.

Direct Subscriber Equipment Segment

Direct satellite subscriber equipment communicates directly with a satellite. This is different than other end-user satellite equipments that receive their communication via a satellite receiving system located in a communication center. Examples of direct satellite communication equipment include; direct broadcast satellite (DBS) TV, the XM radio system, and satellite mobile telephone systems. Typically called subscriber units, consists of an antenna, radio receiver, transmitter (only for two-way systems) and an interface converter (such as a video or audio interface) depending on the application.

There are various types of subscriber units, some of which are intended for general use, and some of which are designed to support specific applications. Subscriber units that are used for mobile or portable voice communication may be capable of several services such as voice, messaging and data. They usually appear very similar to a cellular telephone. Other devices, such as digital television receivers or meter reading devices, are designed specifically for their application.

Broadcast

Broadcast service allows the same signal, such as television or audio signals, to be sent to all the receivers. Satellites used for broadcast services are commonly GEO, allowing fixed antennas dishes on the earth to stay pointing directly at them. These types of satellites are most familiar because they have been around since the 1970s.

Figure 1.3 shows the different types of satellite communication systems. The GEO satellite system is primarily used for television broadcast services, as their satellites appear stationary above the Earth. MEO and LEO systems are used for mobile communications as they are located much closer to the Earth. However, these satellites continuously move relative to the surface of the Earth.

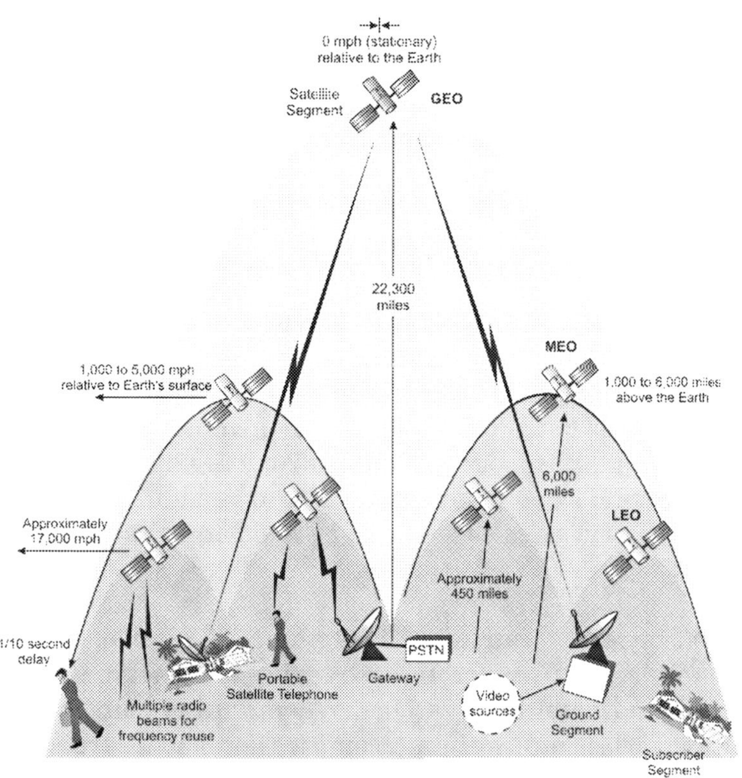

Figure 1.3, Basic Broadcast Satellite System

In a typical broadcast satellite system, there are several program sources (television signals) that are combined into one radio transmission signal. This signal is sent up to a satellite (on the uplink) where it will be re-transmitted back down (the downlink) to a large geographic area.

As satellites of higher frequency are employed, receiving antennas can be made smaller. The smaller "dishes" are less costly and less complicated and become more attractive to larger numbers of customers who are more willing to adopt a satellite service, which uses an 18-inch dish over a 10-meter dish. DBS (direct broadcast satellite) services are able to use smaller dishes because of their operation at higher frequencies.

Digital audio broadcasting (DAB) is another broadcast system that can use satellite systems to deliver high quality digital audio to a large geographic area. Systems are being tested around the world to deliver DAB from satellites as well as from land based (terrestrial) antennas.

Very Small Aperture Terminals (VSAT)

A very small aperture terminal (VSAT) is a ground-based satellite terminal that offer two-way data access to satellite systems. Typical uses for VSAT networks are the broad distribution of data to many receivers. An example of this service is the distribution of price or inventory levels to a chain of stores throughout a large geographic area. They are also used for each store to send inventory and sales information back to headquarters. An important feature of a VSAT service is ease of deployment; installation takes approximately 2 hours.

Because VSAT systems can often communicate in two directions, fast-response protocols are used for time-sensitive transactions such as credit card purchases and hotel reservations. When data is being sent through a VSAT network, it uses efficient data compression for file transfers.

An example of a use for VSAT systems is the sending of new pricing information from a corporate headquarters to its stores located throughout the country. The corporate headquarters provides a data signal to the satellite gateway, which adds the companies' addresses to the message and inserts the data into the uplink satellite data channel. The satellite retransmits the data and the stores, which have a VSAT receiver and the correct address code, will receive it. Some stores may have over 2,000 stores (such as Walmart) that simultaneously receive the message.

Figure 1.4 shows how a very small aperture (VSAT) satellite communication system can be used to efficiently distribute data messages to many receivers in a large geographic region. In this example, a computer at a corporate headquarters building sends data messages to the VSAT communication system via a landline data connection. The VSAT system receives the data message (a corporate price change in this example) and transfers this data message through an Earth station satellite transmitter to the satellite. The satellite receives this data message and retransmits it to the large radio coverage area. Each of the company's retail stores that have the VSAT antenna (a small antenna dish) and receiver decode the message and forward the information to their local computer system which updates the prices.

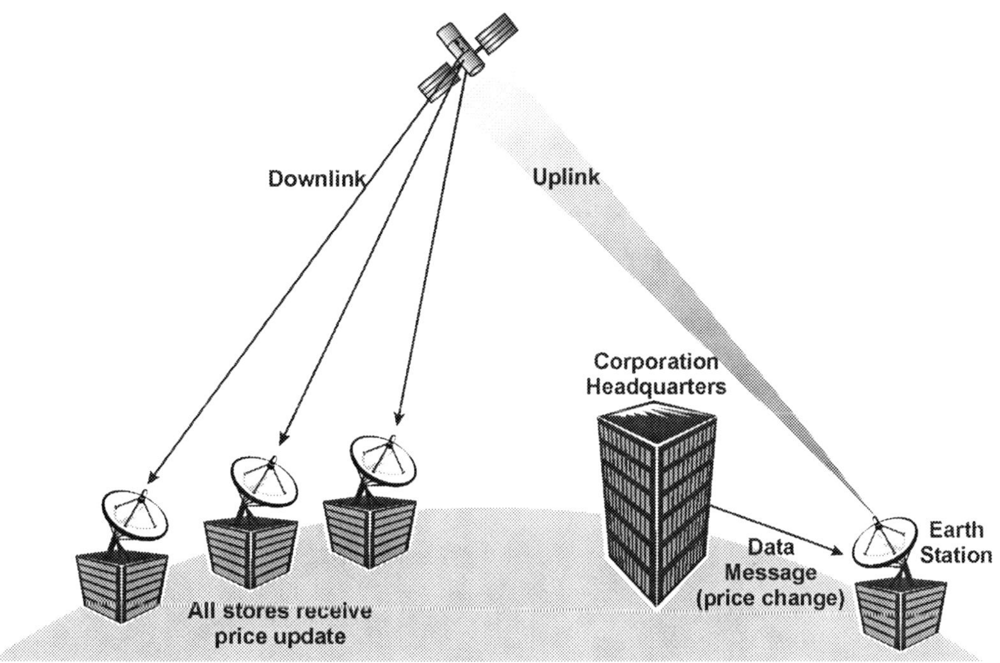

Figure 1.4, VSAT Satellite System

Mobile Satellite Systems (MSS)

Mobile Satellite Systems (MSS) provide two-way communication with mobile satellite radios. MSS systems can be of the GEO, MEO or LEO type. MSS can be divided into maritime mobile satellite services (MMSS), land mobile satellite service (LMSS) and aeronautical mobile satellite services (AMSS).

An example of a mobile satellite system is communication between a ship and a landline telephone system. To establish communication, a shipboard satellite telephone first dials the telephone number of a landline telephone. The satellite radio then sends a request to the satellite for service and the dialed digits are retransmitted (relayed) to the satellite gateway where it can be connected to the public switched telephone network (PSTN). If the service is authorized, the gateway will dial the office telephone and when the call is answered, the audio from the shipboard satellite telephone will be connected to the audio of the office telephone (through the phone system).

Satellite telephones have been relatively large (briefcase size) until recently. Due to the distance of GEO satellites, mobile satellite phones had to use high power amplifiers and relatively large directional antennas. With the introduction of low earth orbit systems, it is possible to use handheld portable satellite telephones. The MEO, and some new GEO systems with very large satellite antennas, will also support handheld portable phones.

Figure 1.5 shows a functional diagram of a portable satellite telephone. This diagram shows that a typical portable satellite telephone consists of a high gain antenna that has multiple antenna elements, antenna control circuitry, radio signal transceiver (transmitter and receiver) and audio signal processing circuits.

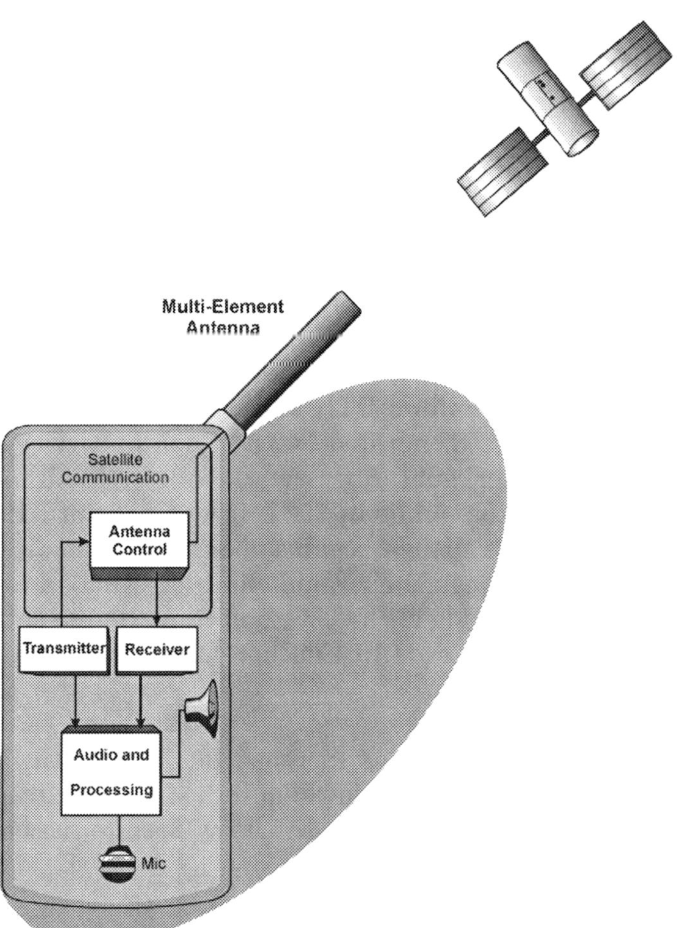

Figure 1.5, Portable Satellite Telephone

Many mobile satellites allow dual mode satellite and terrestrial (ordinarily cellular) service. This service allows the mobile telephone to use the cellular system if it is available. If the cellular system is not available, the mobile telephone will access the satellite system.

Market Growth

In 2002, the worldwide Satellite industry revenue was $86.8 billion worldwide. This was divided into $49.8 billion for services, $21.2 billion for ground equipment manufacturing, $12.1 billion for the manufacturing of satellites, and $3.7 billion for the satellite launch industry [i]. The services provided by satellite systems include video distribution, data distribution (such as Internet connections), mobile satellite service, and satellite radio.

In 2003, there were 61 million DTH customers worldwide and projections indicate there will be 101 million DTH subscribers worldwide by 2008 [ii]. Direct subscription services generated $42.5 billion in service sales in 2002 [iii] with over 80.7 million subscribers worldwide [iv]. Although satellite service was available to the public in 1987, it was not until 1994 when the smaller, compact satellite dishes became available that direct broadcast television receiver sales began their rapid climb. By 1996, a total of 10 million (or 10%) of the households in the U.S. had converted to satellite TV service and by 2002; there were over 19.1 million DBS customers in the United States [v].

Figure 1.6 shows the annual sales increase in the growth in the United States for the DBS/DTH satellite television industry. This graph shows the growth has continued and will continue to grow despite increased competition from cable and DSL services.

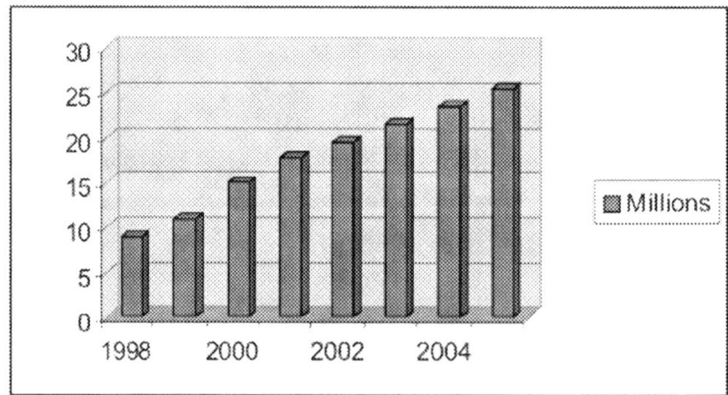

Figure 1.6, US Direct-to-Home Satellite Sales
Source: Satellite News

In mid-2002, Inmarsat, the leading provider of mobile satellite service (MSS) announced that it had activated their 250,000th mobile satellite terminal [vi]. Due to the increased availability of low cost land based mobile communications in the late 1990s (cellular and PCS), optimistic predictions of 12 million mobile satellite telephone and paging devices in service by 2003 have not been realized [vii]. As a result of international events that require global communication, LEO and MEO satellite systems such as Iridium and Globalstar that can use smaller, low cost handsets will likely dramatically increase the number of mobile satellite customers worldwide.

In 1999, VSAT terminal sales exceeded 100,000 units with over $3 billion in service revenue [viii]. By 2002, there were more than 1,000,000 VSAT terminals in use in more than 120 countries [ix]. The VSAT industry is experiencing significant growth in the Internet service provider (ISP) industry segment.

In 2003, the global positioning service (GPS) industry exceeded $13 billion in device sales [x]. One of the key reasons why the GPS industry has grown so quickly is the decline in GPS receiver cost from $3,000 to under $150 per unit [xi]. GPS receivers have also been recently enhanced for increased accuracy to within several feet. This increases the types of applications that can use GPS such as cellular telephones, asset tracking, and vehicle monitoring.

The introduction of paid subscription satellite radio services such as XM and Sirius was the most successful product release in history. Within 60 days of its market introduction at the end of 2001, there were already 30,000 paying customers [xii]. By December of 2004, the leading service provider XM satellite radio already had over 3.1 million paying customers [xiii].

Technologies

Some of the key technologies used in satellite systems include the type of satellite system orbit (GEO, MEO, LEO, or Elliptical), differential position location, multibeam transmission systems, and multiple access (radio channel sharing) systems.

Geosynchronous Earth Orbiting (GEO) Satellites

Geosynchronous earth orbit (GEO) satellites are located approximately 22,300 miles above the surface of the earth. The reason for choosing this particular height above the earth is the precise balance between the gravity pull to the earth and the centrifugal force (speed) of the satellite spinning around the earth, thus the satellite spins with the earth's rotation. It is at this height where the satellite will appear to be fixed in position (stationary) above the Earth because it is spinning at the same rate as the earth (1 revolution per day). In 1963 Hughes Aircraft and NASA achieved the first geosynchronous orbiting satellite.

The primary advantage of a GEO satellite system is that only one satellite is required to provide radio coverage to one-third of the earth's surface. GEO satellites also have an average life span of 10 to 15 years.

Because GEO satellites are located a long distance away, radio transmissions through these satellites have an approximate delay of 1/2 second (1/4 second each way). This transmission delay can result in difficulty when the system is used for voice services.

This figure 1.7 shows a GEO satellite system. This diagram shows that the GEO satellite is connecting a telephone caller on land with a ship. The telephone call is routed through the gateway to the satellite. The satellite transponder converts the frequency and retransmits the signal back to earth. A satellite telephone on the ship receives the radio signal and converts it back to the original audio signal. The diagram also shows that a separate ground control facility is used to monitor and control the position of the satellite.

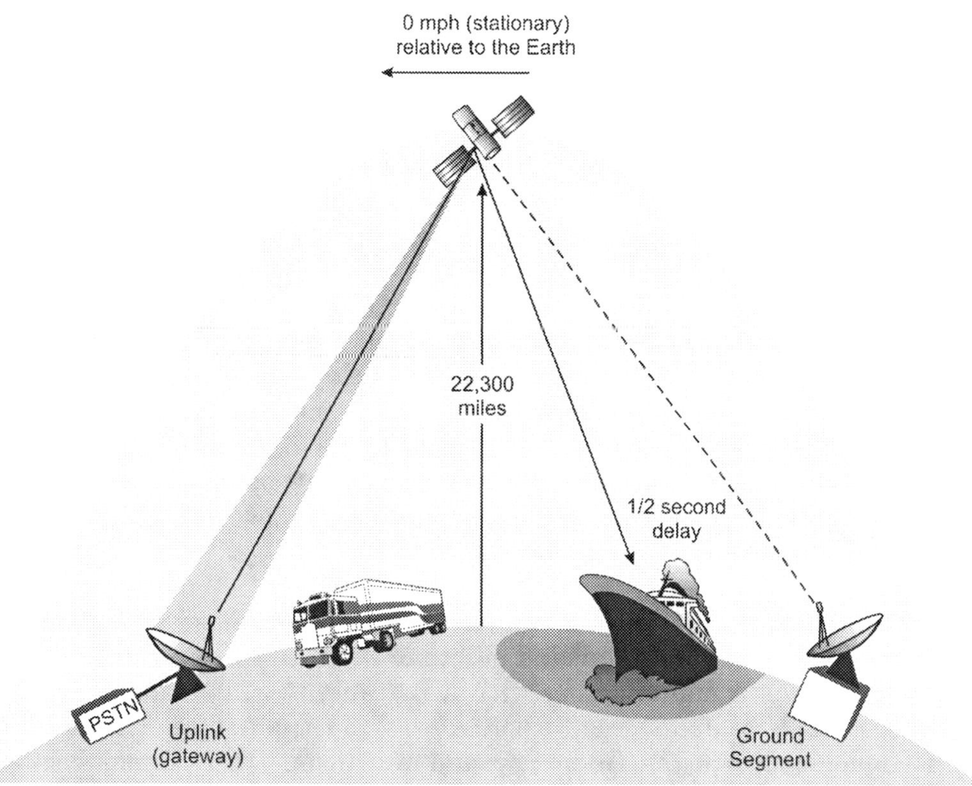

Figure 1.7, GEO Shipboard Communication Satellite System

Medium Earth Orbiting (MEO) Satellites

Medium earth orbit (MEO) satellites are positioned between 1,000 to 6,000 miles above the earth. Unlike GEO satellites, which appear stationary above the earth, MEO satellites move slowly through the sky as seen from an observer on the earth. A typical MEO satellite will circle the earth in approximately 6 hours. Among other benefits, this provides the potential for global coverage if the countries, which the satellites move over, allow radio transmission into their country.

Because the MEO satellites are located closer to the earth, the radio coverage footprint from MEO satellites is somewhat smaller than from GEO satellites. This means 2 to 3 MEO satellites are required to cover the same area as a GEO satellite. Because these satellites move, several MEO satellites must be positioned around the world to allow continuous coverage of a specific area. As one satellite moves out of the radio coverage area, another satellite can take its place. With approximately 10 to 12 satellites, global radio coverage is possible.

Because they are much closer to the earth than GEO satellites, MEO satellites only have about a 1/10th second or less delay time for transmitting a voice/data signal. They are also sometimes of a smaller size than the GEO satellites and therefore they are less expensive to build and launch. Similar to GEO satellites, MEO satellites have an average life span of 10 to 12 years. MEO satellites do allow mobile communications to low to medium power handheld portable satellite telephones.

Figure 1.8 shows a MEO satellite system. In this diagram, several satellites circle the earth at several thousand miles per hour. In this example, a landline telephone is communicating with a portable satellite telephone. The telephone call is routed through the gateway to the satellite. The satellite transponder converts the frequency and retransmits the signal back to earth. The portable satellite telephone receives the radio signal and converts it back to the original audio signal.

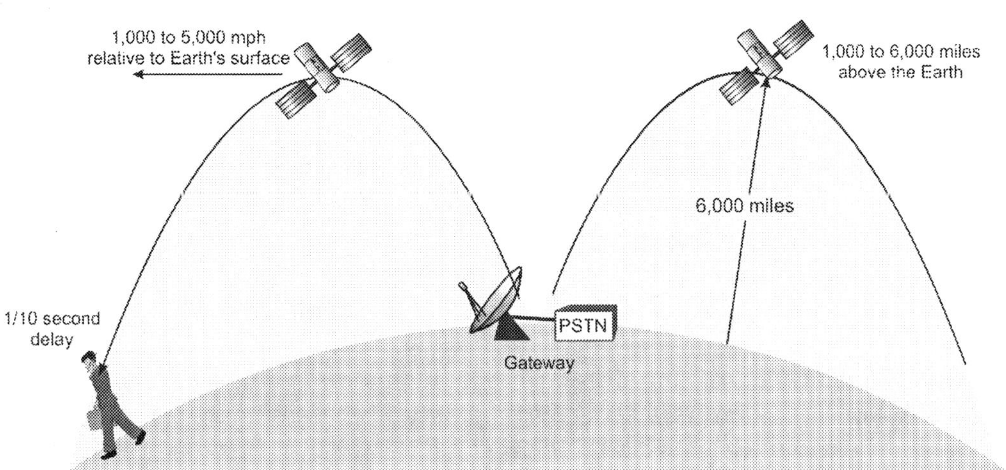

Figure 1.8, MEO Satellite System

Low Earth Orbiting (LEO) Satellites

LEO satellites are located approximately 450-1,000 miles above the Earth. Because LEO satellites are located so close to the Earth's surface, LEO satellite systems normally provide mobile satellite services (MSS) to handheld or mobile satellite telephones.

Also because LEO satellites are located very close to the Earth, each satellite must move at approximately 17,000 miles per hour to avoid falling into the Earth. LEOs circle the earth in approximately 90 minutes. LEO satellites are technically much simpler and more robust than geosynchronous satellites and are less likely to suffer catastrophic failure during deployment or during the satellite lifetime.

The radio coverage footprint of a LEO satellite is small, so 50 or more satellites are required to achieve continuous coverage. As one LEO satellite moves out of the radio coverage area, another satellite can take its place.

Because LEO satellites are only 450 to 1000 miles above the earth (10 to 50 times closer to the earth than MEO or GEO satellites), LEO satellites have a negligible delay time for voice/data signals. LEO satellites are also much smaller and have a much lower cost to build and launch than other types of satellites. They have an average life span of 5 to 8 years, primarily because of radiation effects. LEO satellites do allow mobile communications to low power handheld portable satellite telephones, which do not need highly directional antennas. In practice, the transmit power level can be lower than 1 Watt for portable LEO satellite terminal equipment.

Figure 1.9 shows a LEO satellite system. In this diagram, a portable satellite telephone is communicating with a landline telephone. The satellite telephone communicates with the closest LEO satellite. Because LEO satellites fly very close to the surface of the earth, they go across the visible horizon in approximately 10 minutes in reference to a mobile satellite customer's location. When the first satellite moves out to the horizon, another LEO satellite becomes available to continue the call. However, robust network communications need to be in place to maintain calls (especially data transmission) within this period. Some systems will use satellite diversity to allow talking through more than one satellite at a time, avoiding call "dropouts" from signal blockage.

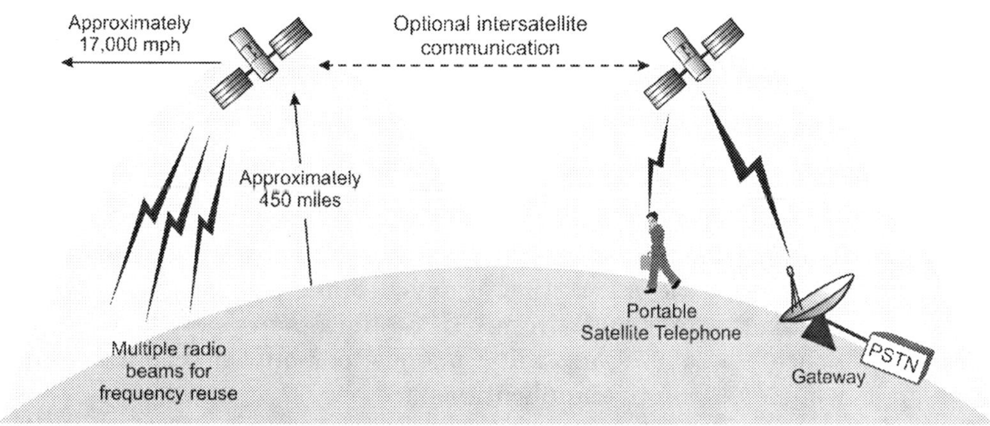

Figure 1.9, LEO Communication Satellite System

Elliptical Orbiting Satellites

Satellite that orbits the Earth where the distance from the surface of the Earth is not constant is called Elliptical orbits. The use of elliptical orbits allows a satellite to operate closer to the Earth for part of its orbit.

A satellite that is in elliptical orbit has an oval-shaped travel path. The part of the orbit that is closest to the center of Earth is called the "perigee". The part of elliptical satellite orbit that is the farthest away is called the "apogee". A satellite in elliptical orbit can take approximately 12 hours to circle the planet.

Most satellites that use this elliptical orbit are exploration satellites with little commercial value not communications satellites. Galileo, Venus and other planet explorers use this orbit to create a "sling shot effect" for launching themselves away from earth to go to Jupiter or something. These are never used for commercial uses. To get to Mars, a satellite will use this elliptical orbit because as it gets closer to earth it speeds up a lot due to the effect of gravity and then slingshots around the earth faster and faster every time it comes back. At a certain point it performs the slingshot and then fires it's rockets and uses all the momentum to get it out of earth orbit towards Mars.

Global Positioning System (GPS)

Global positioning system is a navigation system that uses satellites to act as reference points for the calculation of navigational position. GPS is used extensively by the military and aircraft GPS chipsets are now being incorporated into wireless devices, including phones, personal digital assistants (PDA's), as well as automotive applications.

In the late 1980's, the military began to allow GPS technology to be used for public use. Due to the much larger number of units in production, the price for GPS equipment was dramatically reduced. Because the government maintains the satellites, there is no cost to use GPS service.

There are several systems used for global positioning. These include the Russian Global Navigation Satellite System ("Glonass"), the United States global positioning system (GPS) and another United States positioning system called "Sat-Nav".

The GPS system is based on 24 orbiting satellites called "NAVSTAR." GPS receivers can calculate their exact location (height and location [Longitude and Latitude]) by measuring the distance from four or more satellites. Consumer devices are accurate within one city block, and commercial/military devices can measure to the millimeter (less than 1/16 inch). With the addition of software grid information, a real-time driving location map can be generated and available to those people "who never get lost".

Figure 1.10 shows a global positioning satellite (GPS) system. This diagram shows how a GPS receiver receives and compares the signals from orbiting GPS satellites to determine its geographic position. Using the precise timing signal based on a very accurate clock, the GPS receiver compares these signals from 3 or 4 satellites. Each satellite transmits its exact location along with a timed reference signal. The GPS receiver can use these signals to determine its distance from each of the satellites. Once the position and distance of each satellite is known, the GPS receiver can calculate the position where all these distances cross at the same point. This is the location. This information can be displayed in latitude and longitude form or a computer device can use this information to display the position on a map on a computer display.

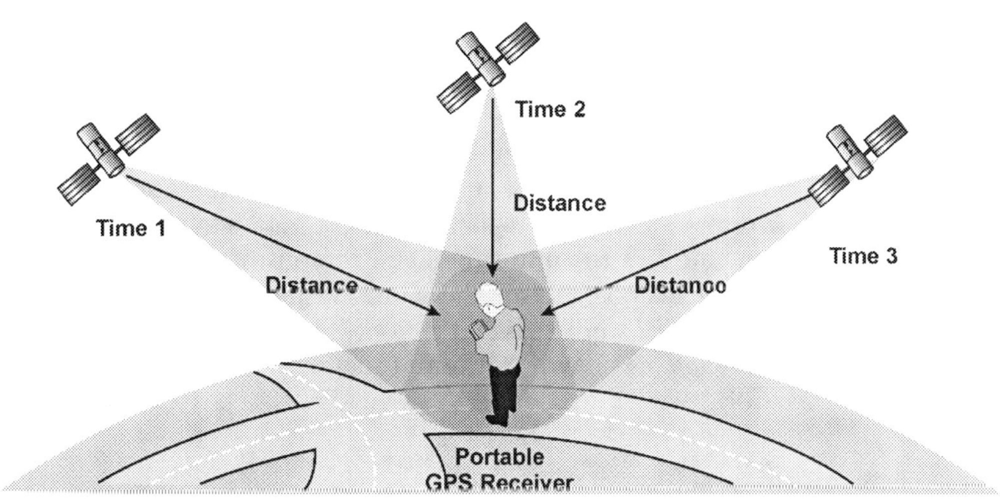

Figure 1.10, Global Positioning System (GPS)

Multibeam Transmission Systems

Multibeam transmission systems allow a single transmitter location to provide multiple communication channels by dividing the radio coverage into focused radio beams that reuse the same frequency. To allow multiple accesses, each mobile radio is assigned to a focused radio beam. These radio beams may be fixed or they may dynamically change with the location of the satellite receiver.

A sophisticated method that is used to provide multibeam transmission is spatial division multiple access (SDMA). SDMA technology allows a satellite to create multiple beams by electrically creating focused energy levels. To create the dynamically changeable beams, the SDMA satellite has multiple antennas that are a phased array. This antenna system can be capable of focusing radio coverage from a few (e.g. 10) too many (over 100) smaller geographic coverage areas within their satellite radio coverage area.

Because a single satellite can provide coverage to an area several thousand miles wide, SDMA technology can use that single satellite to provide several radio beams, some of which reuse the same frequencies. This increases the number of simultaneous channels a single satellite can serve.

Figure 1.11 shows a satellite system that uses spatial division multiple access (SDMA) technology. In this example, a single satellite contains several directional antennas. Some of these antennas use the same frequency. This allows a single satellite to simultaneously communicate to two different satellite receivers that operate on the same frequency. Usually beams that are separated by more than two or three half-power beam widths can use the same frequencies, as shown in the figure.

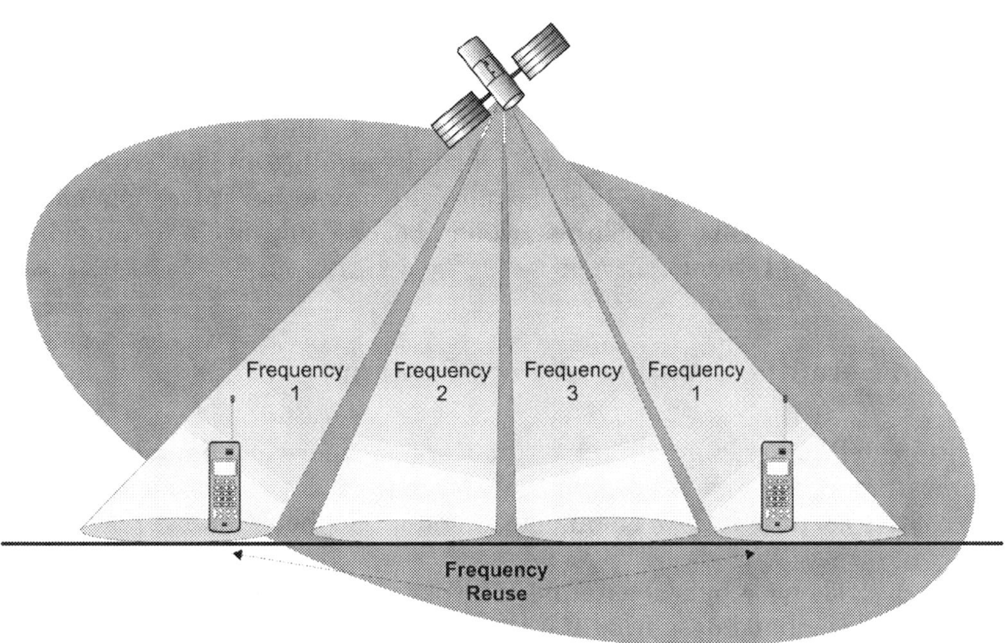

Figure 1.11, Multibeam Satellite Transmission

Multiple Access Technologies

The radio communication systems used by satellite have changed over the years. Some satellite radio channels continue to use analog modulation, however most satellite systems are converting to digital transmission. The method used to share digital channels varies depending on the system and equipment that will be used to access the satellite system. The different types of access technology used for communication satellites include time division multiple access (TDMA), frequency division multiple access (FDMA), code division multiple access (CDMA), and time division duplex (TDD). The access technology types can vary between the uplink and downlink channels for satellites that do on-board processing.

Satellite Cross-Links

You can think of Satellite cross links like an "internet in the sky". Microsoft was considering setting up a network of satellites that would allow users anywhere in the world to connected to a near by satellite and by "cross links" across the network of satellites connect to any other place on earth. Implementing satellite cross-links would be one approach to providing World Wide Web Internet service.

Hybrid Satellite Broadband Data Systems

Hybrid satellite broadband data systems transfer forward broadband communication to the end users and transfer return data from users via an alternative communication channel (e.g., via a telephone line).

Figure 1.12 shows a hybrid satellite Internet system that allows customers to receive broadband data from the Internet via the satellite broadcast network and allows the customer to return data to the Internet via public telephone network. In this example, the satellite receiver set top box contains a connection to the satellite antenna system and it contains a connection to the public telephone network. When the customer desires to access the Internet or download a data file, the customer sends the request via a data modem that is connected to a telephone line. The Internet service provider (ISP) decodes the request and sends the data to the satellite broadcasting company that transfers the broadband data to the user via the satellite.

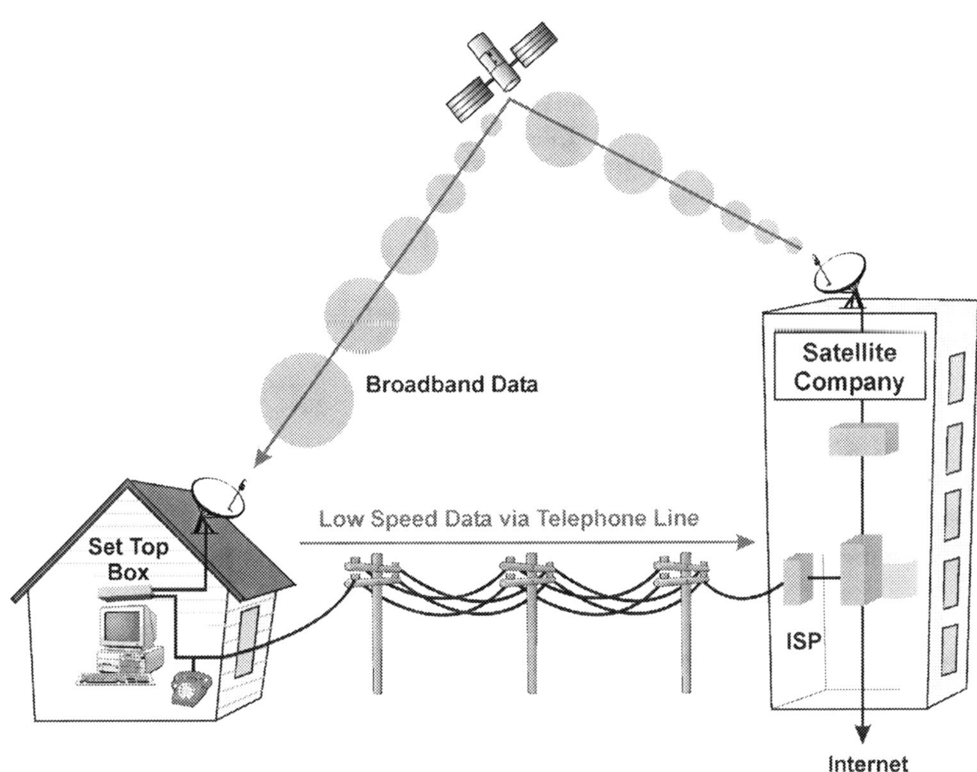

Figure 1.12, Hybrid Satellite Broadband System

Satellite Digital Audio Radio Service (SDARS)

Digital audio radio services (DARS) is a radio system that provides audio programming via satellite transmission. It is similar to Direct Broadcast System (DBS) TV systems. The initial application of DARS was by CD Radio in 1990. It is also called Digital Audio Broadcasting (DAB) outside the U.S. DARS may be provided by land based (terrestrial), satellite, or a combination of satellite and land based repeaters.

Satellite digital audio radio service (SDARS) is a radio broadcasting service that transmits radio signals that carry digital information (such as digital audio and programming information) through satellites to end users. To reach areas that are not accessible to satellite signals (such as tunnels or city areas that block satellite signals), SDARS systems typically use repeaters to retransmit satellite signals to specific geographic areas.

Satellites that provide SDARS use high power transmitters to allow small low cost receivers. Satellite digital radio channels are composed of many digital audio channels. Each digital audio channel has data rates that typically range from 48 kbps to 160 kbps.

The transmission of satellite signals can be blocked and may experience signal fades over time (possibly due to blocked views of satellites). To overcome these channels, satellite radio signals are sent on duplicate frequency bands for frequency diversity and these signals are time delayed to allow one channel to fill in when another channel cannot be received.

Some satellite digital audio systems offer optional Internet connection to allow users to access their digital audio programming via multimedia computers through the Internet.

Figure 1.13 shows how a satellite digital audio radio system transfers audio and multimedia programming via satellites directly to end users (vehicles) and how it may be relayed by earth based (terrestrial) towers to allow the DARS signals to be received in areas that are hard to penetrate by satellite signals. This diagram also shows that there is an optional link through the Internet to end users that allow subscribers to have access to their music channels.

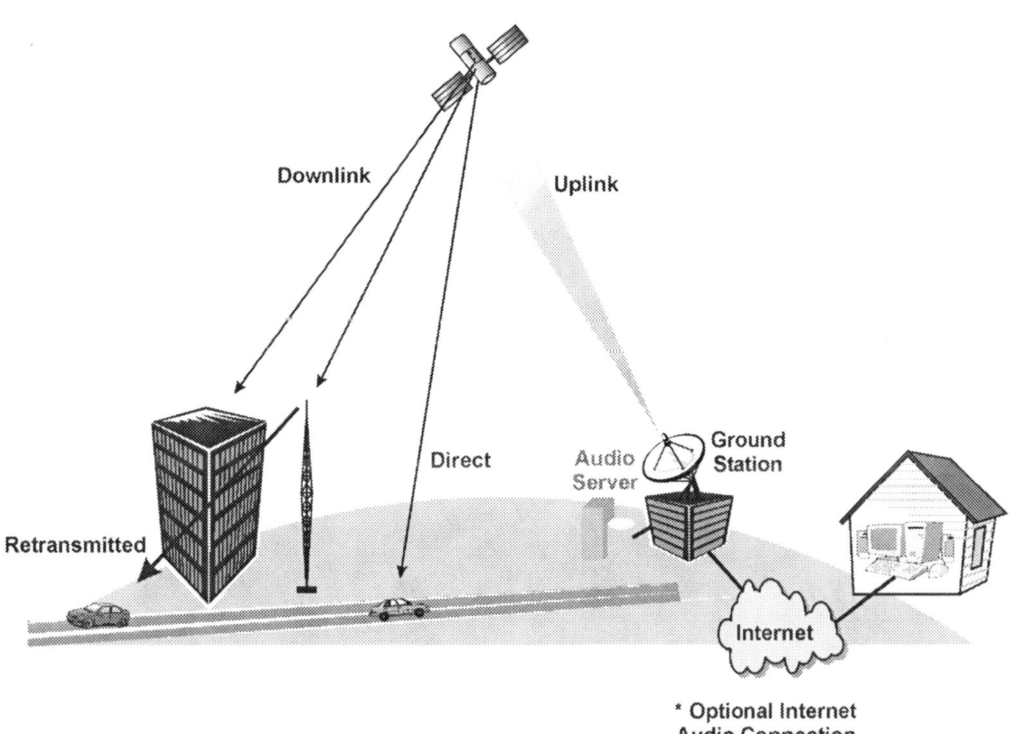

Figure 1.13, Satellite Digital Audio Radio System

Digital Video Broadcasting via Satellite (DVB-S)

Digital video broadcasting is a set of digital television industry standards that are published by the joint technical committee (JTC) of the European Telecommunications Standards Institute (ETSI). These standards can be obtained at www.ETSI.org.

DVB industry specification are grouped by the type of distribution system they use. DVB-C (cable), DVB-DSNG (Digital Satellite News Gathering), DVB-MC and DVB-MS (Multipoint Microwave Video), DVB-S (Satellite), DVT-T (Terrestrial) and others.

A typical DVB-S satellite channel has 36 MHz of bandwidth and it uses quadrature phase shift keying modulation (QPSK) to provide data transmission rates of 35 to 40 Mbps. Each digital television channel is compressed using MPEG-2 compression with a data transmission bandwidth of 4 to 6 Mbps [xiv].

Figure 1.14 shows the basic operation of a satellite digital video broadcasting system. This diagram shows that a DVB system contains several transport streams that carry multiple television and audio channels each. The satellite receiver demodulates the appropriate transport channel and decodes the specific media channel that the user has selected.

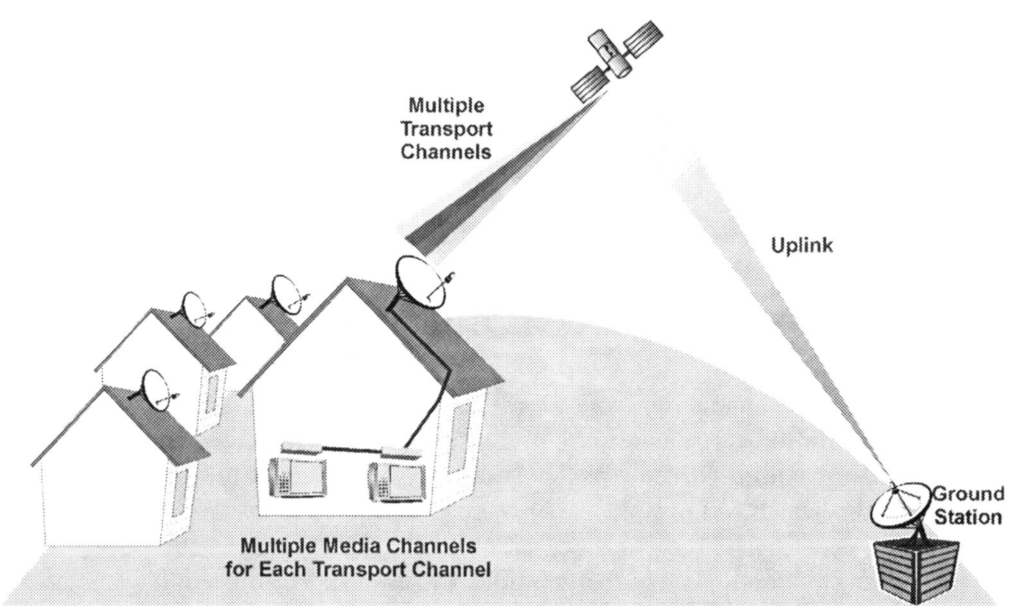

Figure 1.14, Satellite Digital Video Broadcasting System

Digital Video Broadcasting with Return Channel via Satellite (DVB-RCS)

Digital video broadcast with return channel via satellite (DVB-RCS) is an industry standard for high-speed two-way Internet communications via GEO satellite communication. Development of the DVB-RCS specification was started in 1999. The DVB-RCS system defines a return channel satellite terminal (RCST) that is capable of receiving DVB-S digital transport channels along with the ability to retransmit signals (such as channel request information or Internet data) back through the satellite system.

The DVB-RCS system uses multiple frequency time division multiple access (MF-TDMA) to share the communication channels between the RCST and the satellite [xv]. RCSTs are managed by a satellite system network control centre (NCC). The NCC sends signaling tables to the MPEG media streams to identify the different types of programs (television channels, audio, or data). The NCC provides access control information (allowable burst transmission time intervals) to the NCST to help coordinate multiple access attempts to the reverse channel.

There are some challenges when using GEO and MEO satellite links to connect users the Internet because of the added time delay to packet data transmission. Standard Internet protocols to automatically change the packet size to smaller packets due to the long transmission time and the timeout of communication sessions. These challenges can be overcome by using a satellite data transmission gateway can adapt and adjust the protocols as the pass between the RCST and the Internet.

Figure 1.15 shows how a satellite system can provide two-way communication capabilities. This diagram shows that the forward direction (satellite to end user) has higher data transmission rates than is available from the user to the satellite. This example also shows that the broadband satellite communication system includes an Internet gateway that can optimize the IP communication session to compensate for challenges created by the added transmission delay time.

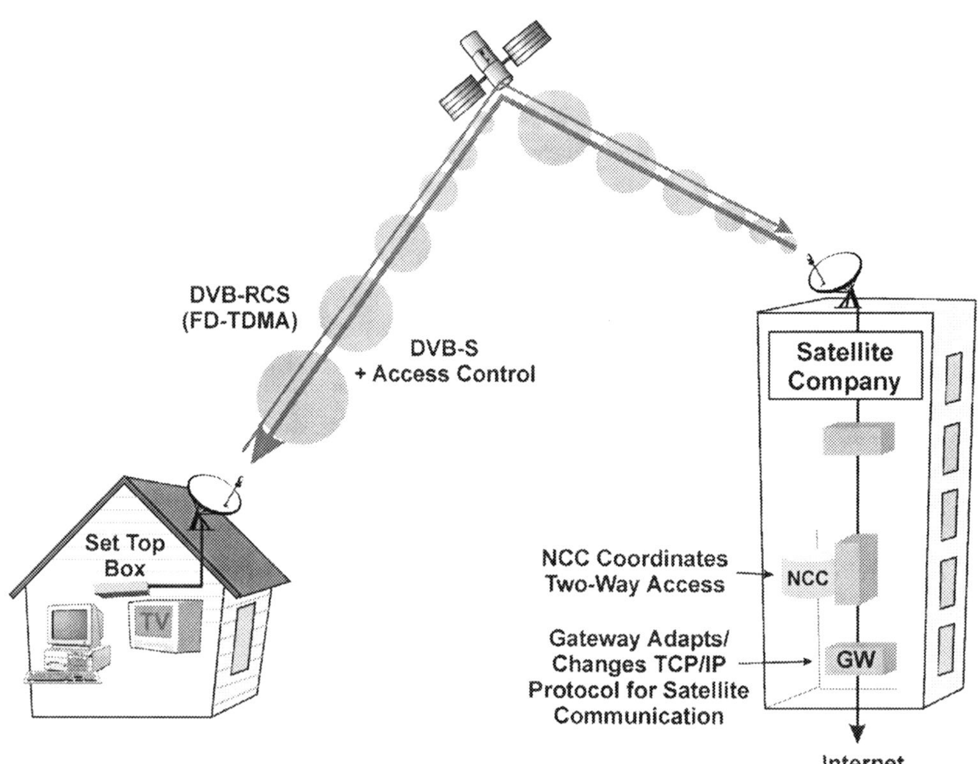

Figure 1.15, Two-Way Satellite Digital Video Broadcast System (DVB-RCS)

Bent Pipe Satellite

Bent pipe satellite transmission is a process of transferring radio signals through a satellite by redirecting the signal (bending) to a new direction rather than receiving, processing, and re-transmitting the signal. The bent pipe method reduces the complexity of the transferring device as it is only retransmits the original signal in a new direction without analyzing or altering its information content.

MPEG

Motion pictures expert group is a working committee that defines and develops industry standards for digital video systems. These standards specify the data compression and decompression processes and how they are delivered on digital broadcast systems. MPEG is part of International Standards Organization (ISO).

MPEG technology (or variants of MPEG technology) are used to compress the digital video signal and combine the digital signal with audio and other information about the media program. The most common MPEG technology used in digital television is MPEG-2. MPEG-2 is a digital video encoding process that compresses digital video up to 200 to 1. MPEG-2 is the current choice of video compression for digital television broadcasters as it can provide digital video quality that is similar to NTSC with a data rate of approximately 3.8 Mbps.

Figure 1.16 shows how MPEG transmission can be used to combine video, audio, and data onto one packet data communication channel. This example shows that multiple types of signals are digitized and converted into a format suitable for the MPEG packetizers. This example shows a MPEG channel that includes video, audio, and user data for a television message. This example shows that each media source is packetized and sent to a multiplexer that combines the channels into a single transport stream. The multiplexer also combines program specific information that describes the content and format of the media channels. The multiplexer uses a clock to time stamp the MPEG information to allow it to be separated and recreated in the correct time sequence.

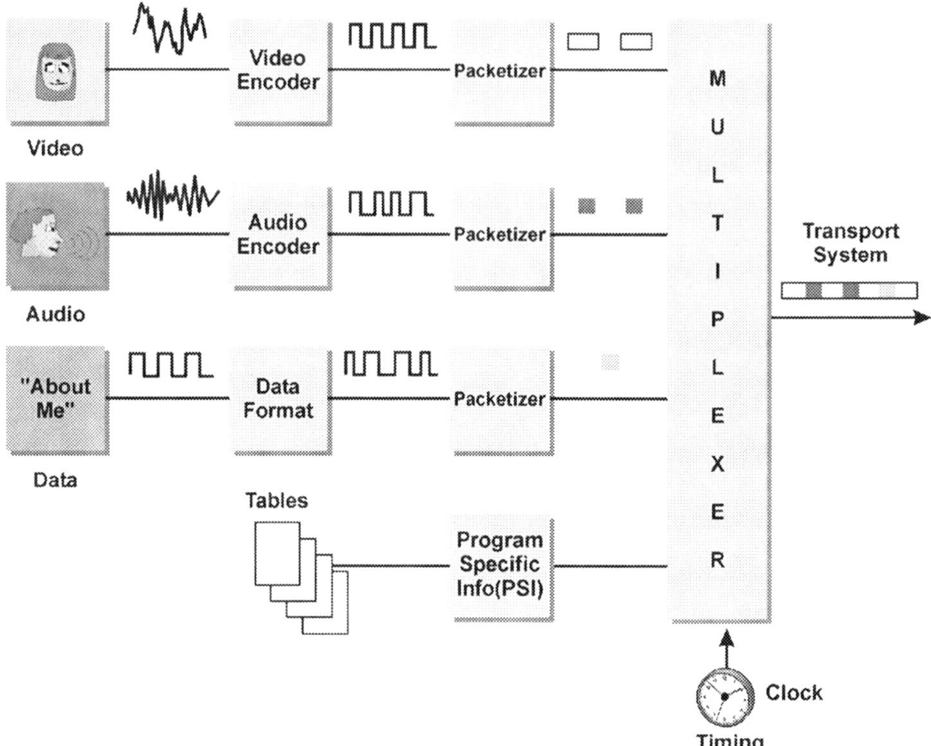

Figure 1.16, MPEG Channel Multiplexing

Mobile Satellite Systems

There are hundreds of satellite systems that are in commercial use and several more that are in final testing or nearing commercial deployment. The following list includes many of the satellite systems that are in use in the Americas.

INMARSAT

Inmarsat business traditionally are maritime, aeronautical, land-mobile and remote-area markets with satellite service from Geosynchronous satellites. They are now moving into the convergence of information technology, telecom and mobility with their new Inmarsat I-4 satellite system, which from 2005 will support the Inmarsat Broadband Global Area Network (B-GAN) - mobile data communications at up to 432kbit/s for Internet access, mobile multimedia.

G2

G2 Satellite Solutions provides end-to-end satellite communications (SATCOM) services to government entities, both domestically and internationally, as well as to certain private sector customers. G2 responds to user requirements (technical, financial and contractual) by tailoring unique solutions from commercially available satellite products and services. As a value-added reseller representing an extensive array of satellite bandwidth and terminal providers, G2 is able to offer "one stop shopping" for integrated, worldwide satellite communications. The G2 Operations Center provides a point of contact for ongoing customer support needs and satellite network monitoring services.

PanAmSat providing global video and data broadcasting services via satellite owns G2 Satellite Solutions. It was part of the Hughes companies, which include Space and Communications Company, Hughes Network Systems (HNS) and DIRECTV. Hughes is also a major shareholder in American Mobile Satellite Corporation, the nation's largest mobile satellite services provider.

Intelsat offers telephony, corporate network, video and Internet solutions around the globe via capacity on 25 geosynchronous satellites in prime orbital locations. Customers in approximately 200 countries rely on Intelsat satellites and ground resources for quality connections, global reach and reliability.

Iridium

Iridium is a LEO satellite system designed primarily for worldwide mobile telephone coverage. The Iridium system has 66 satellites that are used to provide global radio coverage.

Each Iridium satellite projects a grid of beams (cells) onto the Earth's surface in a manner similar to, but larger than, the grid of cell sites in a cellular telephone network. This allows for frequency reuse and much higher system capacity.

A unique attribute of the Iridium system is that each satellite has the ability to directly communicate with adjacent satellites ("cross-link"). This permits calls to be connected from satellite to satellite without the assistance of a ground station. The Iridium system has the capability of switching calls directly through the satellite network. The first Iridium satellites were launched in 1997 and Iridium began commercial operation in 2001.

Globalstar

Globalstar is a MEO satellite system that uses a fairly simple "bent pipe" approach that simply amplifies, upconverts or downconverts, and reflects the radio signals back to earth. This approach smaller, less complex satellites with all the call processing and switching operations performed on the ground. Each satellite has directional antennas to create many individual radio coverage areas (spot beams). This increases the system capacity and better controls the radio coverage area. The first Globalstar satellites were launched in 1998 and commercial service began in February 2000.

ICO Global Communications

ICO Global Communications is a MEO satellite system that is designed to provide communication to mobile and fixed users around the world. ICO is planning to offer voice, data, and wireless Internet services.

Teledesic

Teledesic is a "Big LEO" satellite system that was designed to provide high-speed data services (broadband) to consumers. The term "Big LEO" is used to indicate in excess of 200 satellites. The Teledesic system was a $9 billion project initiated by Bill Gates and Craig McCaw. Teledesic was planned to become available in the early 2000s. However, in 2002, Teledesic announced it had discontinued development and returned its frequency licenses to the FCC and ITU.

The Teledesic system is designed to transmit data from 16 Kbps to 2 Mbps inexpensively with high quality and low delay. This will allow Teledesic to provide for both voice and data services. Teledesic is primarily expected to provide affordable high speed Internet access to remote areas. Teledesic has planned for 288 satellites in 21 different levels (orbital planes). At any time, each Teledesic subscriber unit will be able to connect to at least two of them from anywhere in the world. Teledesic satellites have the capability to communicate with adjacent satellites at speeds of 155 Mbps. Teledesic has recently formed business arrangements with Boeing and Motorola for design and construction of the satellites. (Teledesic changed from 840 to 288 satellites shortly after joining forces with Boeing.)

Digital Audio Satellite Systems

As of 2005, there were 3 key satellite digital audio radio service providers; XM Radio, Sirius, and Worldspace.

Sirius

Sirius satellite radio is a satellite digital audio broadcasting company that uses 3 elliptical orbiting satellites ("moving position") that are used to provide SDARS service to the United States. The frequency band used by the Sirius radio system is 2320.0 MHz to 2332.5 MHz (12.5 MHz of bandwidth).

XM Radio

XM satellite radio is a satellite digital audio broadcasting company that uses 2 GEO ("stationary position") satellites that are positioned over the east coast and west coast of the United States. The satellite in the western half is called "Rock" and the satellite in the eastern half is called "Roll." Each satellite has approximately 3,000 watts of transmitted power with a gross data transmission rate of 3.28 Mbps per carrier. The frequency band used by the XM radio system is 2332.5 MHz to 2345 MHz (12.5 MHz of bandwidth).

WorldSpace

Worldspace is a satellite digtial audio broadcasting (DAB) company that provides services in Asia, Africa, and South America. The frequency band used by the Worldspace radio system is 1452 MHz to 1492 MHz (40 MHz of bandwidth).

Services

Satellite services include long term leasing of satellite channels, wide area dispatch services, voice and data services. The pricing of satellite service is based on many variables, including the availability and cost of substitute services, the cost of providing service and the nature of the customer application. Pricing generally is based on a wholesale pricing structure that incorporates an initial activation charge, a recurring monthly charge for access to the system and charges based on the customer's usage. Additional pricing considerations include access priority and messaging transport usage.

As competition increases in the satellite industry, it is likely that multiple pricing alternatives will be offered including peak/off-peak, volume discounts and annual contract commitment options. Because the satellite industry uses many resellers of service, the usage pricing for services will be largely outside the control of the satellite system owner and will be established by value added resellers (VARs).

Transponder Leasing

It is common that a portion of the transponder such as a single radio channel or portions of a transponder's radio channel may be leased out for the life of the satellite to a value added reseller (VAR). The VAR commonly subdivides the transponder channel to provide temporary communication services to their clients that include video links for news services, teleconferencing, virtual local area networks (VLANS), VSAT, position tracking service, global voice and messaging communications.

The typical lease for a transponder can be several hundred thousand dollars per month depending on the capacity of the channel (e.g. 6 MHz), coverage area and other system services that are provided by the satellite owner.

Dispatch

Satellite systems allow for cost effective wide area (typically nationwide) dispatch services. Dispatch services allow a single user to communicate with one to many users simultaneously in a customer-defined group using a push-to-talk device. Dispatch service can be used to coordinate operations between workers that are operating over wide and/or remote areas.

Companies that benefit from dispatch services include oil and gas pipeline companies, utilities and telecommunications maintenance fleets, state and local public safety organizations and public service organizations with a requirement to communicate on a nationwide basis.

The service fees for dispatch type services vary from a fixed monthly fee with an almost unlimited amount of service to per minute charges ranging up to $7.50 per minute.

Telephony Voice

Satellite telephones are usually capable of two-way voice, facsimile and data services. In the early 1990's, the hardware cost for a mobile satellite telephones ranged from approximately $4,000 to $10,000 each. With the introduction of LEO satellite systems that can use low power transmitters, handheld satellite telephones are expected to cost $750 or below.

Users of satellite telephone voice services are usually charged a fixed monthly access fee and variable usage fees. Monthly bills for satellite voice customers range from under $50 per month for occasional users to over $400 per month for high usage customers. One of the most well known forms of satellite telephone service is an international airphone service. Providers of airphone service include COMSAT and INTELSAT.

Figure 1.17 shows some mobile sample satellite rate plans for different types of satellite systems. This table shows that the per-minute charges for mobile satellite telephone service ranges from 55 cents per minute to $1.87 per minute. In addition to the per minute usage fee, there is usually a minimum monthly service subscription fee. To get the lower usage rates, a higher monthly subscription fee (up to $249.95) is required.

	Iridium		Globalstar		Inmarsat
Monthly	$34.99	$269.99	$34.95	$249.95	$25
Per Min	$1.40	$1.00	$0.99	$0.99	$0.72-$1.87

Figure 1.17, Mobile Satellite Rate Plans in 2003

Data

Satellite data service users are usually charged a monthly access fee that includes a fixed increment of data usage that may consist of vehicle location reports or data messages. Usage beyond the fixed increment is metered and charged on a variable basis depending on the length and mode of transmissions. Satellite data services offer a wide variety of volume packaging and discounts, consistent with the demands of the targeted markets.

Basic data services include fixed and mobile asset tracking, monitoring and messaging services. Primary fixed asset-monitoring applications include electric utility meters, oil and gas storage tanks and wells, oil and gas pipelines and environmental monitoring. Mobile asset tracking includes tracking the location and status of commercial vehicles, trailers, shipping containers, rail cars, and heavy equipment, fishing vessels and barges and government assets. Messaging services include short alphanumeric paging-like communications services and wireless email.

The cost for data transmission via satellites has dropped from approximately 1 cent per byte of information ($10 per kilobyte) to under 0.1 cent per byte ($1 per kilobyte). It is likely the cost per kilobyte will continue to drop as more satellite systems offer low-speed and high-speed data transmission services.

Digital Audio Broadcasting (DAB)

Satellite digital audio radio broadcasting services (SDARS) are usually provided on a monthly subscription fee basis. Because there is a subscription fee to pay for the service, many of the audio channels do not contain advertising messages. Because the system uses digital transmission, satellite digital audio systems usually include information services such as life traffic and weather updates.

There are approximately 150 audio channels available for Worldspace and 100 audio channels for XM Radio and Sirius. The audio channels are a mix of commercial free music, news, weather, entertainment, and information channels. XM Radio and Sirius have the option for listening to the audio channels on a multimedia computer. Along with the audio information, additional information is sent about the media that is being played such as the title and artist of the music.

Because of the channel coding (and the need of the service providers to collect money for their services), DAB radios must be activated for use. Activation fees (if charged) vary based on the method of activation either by telephone or via the Internet. Activation of service through the Internet is typically less expensive.

Figure 1.18 shows some of the different rate plans for satellite digital audio radio service for XM radio, Sirius and Worldspace. This table shows that the basic monthly fee for satellite digital audio service ranges from $5.00 per month to $12.99 per month. This table also shows that listeners may also be able to receive their satellite audio programming via the Internet to their computer (online listening).

	XM Radio	Sirus	WorldSpace
Monthly Fee	**$9.99**	**$12.99**	**$9.99 (Afristar) $5.00 (Middle East)**
Online Listening	**$3.99**	**Free**	**N/A**
Activation Fee	**$14.99 by phone $9.99 by Internet**	**$15.00 by phone $10.00 by Internet**	**None**

Figure 1.18, Satellite Digital Audio Rate Plans in 2005

Digital Video Broadcasting (DVB)

Digital video broadcasting (DVB) services is the providing of digital video and multimedia signals direct to end users who have satellite receivers.

Basic digital video packages typically include approximately several hundred channels of video and audio programming. Of these, 100 to 200 are nationwide television channels and some local channels. The providing of local channels was an issue in the United States until congress passed a mandatory ruling in 1999 that force local television stations to allow satellite companies to retransmit their television signals. In addition to basic channels, additional premium services may be provided which include sports channels and movie channels.

The average revenue per customer for DBS/DVB services was approximately $63 per month in the USA [xvi] and £31.42 per month in the UK [xvii]. Of this, satellite broadcasters paid over 52% of the total revenues to content provides (e.g. television networks, movies, and sports coverage rights) [xviii].

Figure 1.19 shows a comparison of the rate structure for the leading DBS/DVB service providers. This table shows that basic DBS/DVB service start at approximately $25 per month and can exceed $100 per month with the addition of premium (e.g. sport channels) and other media channels (e.g. movie channels).

	DirecTV (Americas)	Dish Network (Americas) (EchoStar)	Sky TV (UK)
Basic	$36.99/mo.	$24.99/mo.	
Premium	$87.99/mo.	$82.99/mo.	
Movies	$12.00/mo. or less	$11.99/mo.	
Local Channels	Add $3.00	Add $5.00	N/A

Figure 1.19, Satellite Digital Video Rate Plans in 2005

Future Enhancements

Satellite technology is changing rapidly. It is likely that future advances will include more efficient satellite radio transmission, balloon antennas and airborne sky stations.

More Efficient Satellite Radio Transmission

Satellite technology is evolving to allow higher speed digital services through the use of LEO systems, SDMA technology and digital access technology. Because LEO systems have many satellites, their ability to reuse frequencies often increases the total data rate capacity of the system. Because each of the new LEO satellites uses multi-beam antennas that permit frequency reuse, this also increases the total data capacity of the system. More efficient digital transmission technologies (more bits per Hertz) are continuing to be developed. The combination of these key technologies and others will allow user data rates to exceed 2 Mbps.

Balloon Antenna System

A balloon antenna system is a communication system that uses balloons to elevate communication equipment. Balloon based communication systems use wireless repeaters (combined receiver/transmitters) to efficiently provide radio coverage to relatively large geographic areas.

Airborne Communication Platforms

Another future satellite technology is the use of airborne sky stations. These airborne platforms will have an antenna the size of a football field that will be located just miles above the surface of the Earth. With the large antenna size and small distance from the surface of the Earth, radio signals transmitted from low power handheld units can be handled efficiently.

One company, Sky Station International, has filed applications with the Federal Communications Commission proposing this new global wireless communications system. Using a network of 250 stratospheric balloons (dirigibles), Sky Station plans to offer wireless services to more than 80 percent of the world's population by 2002. Each lighter-than-air dirigible will comprise two blimp-like aircraft to suspend the large antenna between them. Once completed, the network (which reportedly costs $4.2 billion) will provide 64-Kbps digital broadband service to 1.5 billion customers worldwide. The network will provide services such as wireless Web access and picture-phone connectivity. Sky Station has asked the FCC to designate frequencies at 47 GHz for the system's operation in the U.S.

[i]. *"Satellite Industry Statistics 2002,"* Satellite Industry Association, Prepared by Futron Corporation, www.SIA.org.

[ii]. "Digital Satellite TV Platforms Continue to Gain Subscriber, and Profits are on the Rise," In-Stat, December 7, 2004..

[iii]. *"Satellite Industry Statistics 2002,"* Satellite Industry Association, Prepared by Futron Corporation, www.SIA.org, pg 9.

[iv]. *"Facts and Figures,"* Satellite Broadcasting and Communications Association, www.sbca.com.

[v]. *"2003 Marks Dramatic Shift for DBS,"* Jimmy Schaeffler, Satellite News, January 27, 2003, www.satellitetoday.com.

[vi]. "Inmarsat continues to demonstrate market leadership with 250,000th mobile customer," Press Release, September 18, 2002, Inmarsat, www.Inmarsat.com.

[vii]. "Market Demand for Mobile Satellite Services: Satellite Statistics," Leslie A. Taylor, February 4th, 1999.

[viii]. *"VSAT Markets and Forecasts 2000,"* Comsys, www.comsys.co.uk.

[ix]. *"Multilink Solutions via Satellite,"* European Telecommunications Standards Institute, www.ETSI.org.

[x]. Industrial Economics and Knowledge Center (IEK) via www.itfacts.biz, September 17, 2004.

[xi]. *"GPS now found in everyday products,"* May Wong, Associated Press, posted on www.philly.com on Mar. 28, 2002.

[xii]. "XM Ranks as Fastest-Selling Audio Product in 20 Years with Extremely High Customer Satisfaction," Press Release, XM Radio, January 7, 2002, www.XMRadio.com.

[xiii]. "XM Satellite Radio tops 3.1 million subscribers ahead of year end goal," Press Release, XM Radio, 27 December, 2004, www.xmradio.com.

[xiv]. "DVB Transmission," Dr. Gorry Fairhurst, University of Aberdeen, UK, Jan 2001.

[xv]. "DVB Transmission," Dr. Gorry Fairhurst, University of Aberdeen, UK, Jan 2001.

[xvi]. SEC Form 10k, DirectTV Goup Inc, March 17, 2004, www.SEC.gov.

[xvii]. "Digital Television Update," Q3 2004, 14, Office of Communications (OfCom), December 2004.

[xviii]. SEC Form 10k, DirectTV Goup Inc, March 17, 2004, www.SEC.gov.

Index

Wireless Books

by ALTHOS Publishing

Wireless Systems

ISBN: 0-9728053-4-6 Price: $34.99
Authors: Lawrence Harte, Dave Bowler, Avi Ofrane, Ben Levitan
#Pages 360 Copyright Year: 2004

Wireless Systems; Cellular, PCS, 3G Wireless, LMR, Paging, Mobile Data, WLAN, and Satellite explains how wireless telecommunications systems and services work. There are many different types of wireless systems competing to offer similar types of voice, data, and multimedia services. This book describes what the functional parts of these systems are and the basics of how these systems operate. With this knowledge,

Wireless Dictionary

ISBN: 0-9746943-1-2 Price: $39.99
Author: Althos #Pages: 628 Copyright Year: 2004

The Wireless Dictionary is the Leading wireless industry resource. The Wireless Dictionary provides definitions and illustrations covering the latest voice, data, and multimedia services and provides the understanding needed to communicate with others in the wireless industry. This book is the perfect solution for those involved or interested in the operation of wireless devices, networks, and service providers.

Introduction to 802.11 Wireless LAN (WLAN)

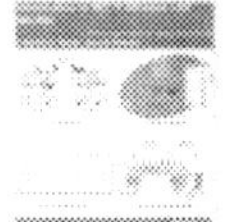

ISBN: 0-9746943-4-7 Price: $14.99
Author: Lawrence Harte #Pages: 52 Copyright Year: 2004

Introduction to 802.11 Wireless LAN (WLAN), Technology, Installation, Setup, and Security book explains the functional parts of a Wireless LAN system and their basic operation. You will learn how WLANs can use access points to connect to each other or how they can directly connect between two computers. Explained is the basic operation of WLAN systems and how the performance may vary based on a variety of controllable and uncontrollable events. This book will explain the key differences between the WLAN systems.

Introduction To Wireless Systems

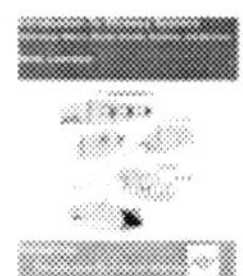

ISBN: 0-9742787-9-3 Price: $11.99
Author: Lawrence Harte, #Pages: 68 Copyright Year: 2003

Introduction to Wireless Systems book explains the different types of wireless technologies and systems, the basics of how they operate, the different types of wireless voice, data and broadcast services, key commercial systems, and typical revenues/costs of these services. Wireless technologies, systems, and services have dramatically changed over the past 5 years. New technology capabilities and limited restrictions are allowing existing systems to offer new services.

Althos Publishing, 404 Wake Chapel Road, Fuquay NC 27526 USA
1-919-557-2260 1-800-227-9681 Fax 1-919-557-2261 WWW.AlthosBooks.com

Introduction to Paging Systems

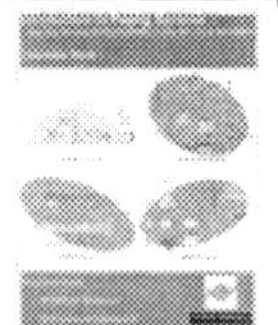
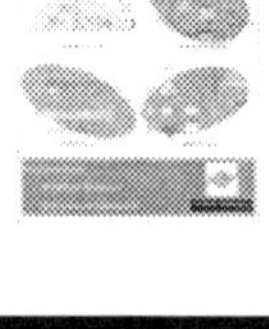
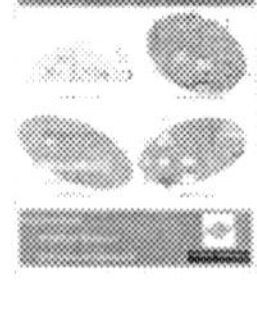
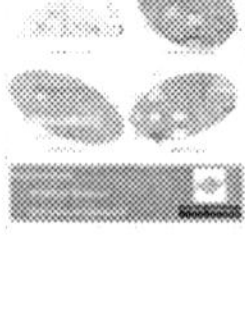

ISBN: 0-9746943-7-1 Price: $14.99
Author: Lawrence Harte #Pages: 48 Copyright Year: 2004

Introduction to Paging Systems describes the different types of paging systems, what services they can provide, and how they are changing to meet new types of uses. This book explains the different types of paging systems and how they are changing. Explained is how and why paging systems are transitioning from one-way systems to two-way systems.

Introduction to Satellite Systems

ISBN: 0-9742787-8-5 Price: $11.99
Author: Ben levitan, Lawrence Harte #Pages: 48 Copyright Year: 2004

In 2003, the satellite industry was a high-growth business that achieved over $83 billion in annual revenue. This book offers an introduction to existing and soon to be released satellite communication technologies and services. It covers how satellite systems are changing, growth in key satellite markets and key technologies that are used in satellite systems.

Introduction to Mobile Data

ISBN: 0-9746943-9-8 Price: $14.99
Author: Lawrence Harte #Pages: 628 Copyright Year: 2004

Introduction to Mobile Data explains how people use devices that can send data via wireless connections, what systems are available for providing mobile data service, and the services these systems can offer. This book explains the basics of circuit switched and packet data via wireless mobile systems. Included are descriptions of various public and private systems that are used for data and messaging services.

Introduction to Private Land Mobile Radio

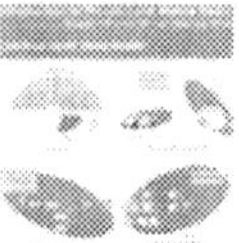

ISBN: 0-9746943-6-3 Price: $14.99
Author: Lawrence Harte #Pages: 50 Copyright Year: 2004

Introduction to Private Land Mobile Radio explains the different types of private land mobile radio systems, their basic operation, and the services they can provide. This book covers the basics of private land mobile radio systems including traditional dispatch, analog trunked radio, logic trunked radio (LTR), and advanced digital land mobile radio systems. Described are the basics of LMR technologies including simplex,and half-duplex.

Introduction to GSM Systems

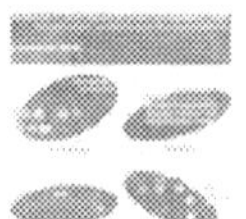

ISBN: 1-9328130-4-7 Price: $14.99
Author: Lawrence Harte #Pages: 48 Copyright Year: 2004

Introduction to GSM describes the fundamental components, key radio and logical channel structures, and the basic operation of the GSM system. This book explains the basic technical components and operation of GSM technology. You will learn the physical radio channel structures of the GSM system along with the basic frame and slot structures.

Introduction to Private Telephone Networks 2nd Edition

ISBN: 0-9742787-2-6 Price: $12.99
Author: Lawrence Harte #Pages: 48 Copyright Year: 2004

Private telephone networks are communication systems that are owned, leased or operated by the companies that use these systems. They primarily allow the interconnection of multiple telephones within the private network with each other and provide for the sharing of telephone lines from a public telephone network.

Introduction to Telecom Billing

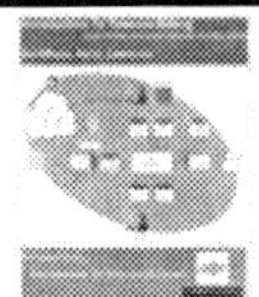

ISBN: 0-9742787-4-2 Price: $11.99
Author: Lawrence Harte #Pages: 36 Copyright Year: 2003

This book explains how companies bill for telephone and data services, information services, and non-communication products and services. Billing and customer care systems convert the bits and bytes of digital information within a network into the money that will be received by the service provider. To accomplish this, these systems provide account activation and tracking, service feature selection, selection of billing rates for specific calls, invoice creation, payment entry and management of communication with the customer.

Introduction to Public Switched Telephone Networks 2nd Edition

ISBN: 0-9742787-6-9 Price: $34.99
Author: Lawrence Harte #Pages: 48 Copyright Year: 2004

Public telephone networks are unrestricted dialing telephone networks that are available for public use to interconnect communications devices. There are also descriptions of many related topics, including: Local loops, switching systems, numbering plans, market growth, public telephone system interconnections, and common channel signaling (SS7),

Introduction to SS7 & IP Telephony

ISBN: 0-9746943-0-4 Price: $14.99
Author: Lawrence Harte #Pages: Copyright Year: 2004

The Introduction to Signaling System 7 (SS7) and IP control system that is used in public switched telephone networks (PSTN) can be interconnected to other types of systems and networks using Internet Protocol (IP). Some of the interconnection issues relate to how the control of devices can be performed using dissimilar systems.

Introduction to IP Telephony

ISBN: 0-974278-7-7 Price: $12.99
Author: Lawrence Harte #Pages: Copyright Year: 2003

This "Introduction to IP Telephony" book explains why companies are converting some or all of their telephone systems from dedicated telephone systems (such as PBX) to more standard IP telephony systems. These conversions allow for telephone bill cost reduction, increased ability to control telephone services, and the addition of new telephone information

Signaling System Seven (SS7) Basics 3rd Edition

ISBN: 0-9728053-7-0 Price: $34.99
Author: Lawrence Harte #Pages: 276 Copyright Year: 2003
This introductory book explains the operation of the signaling system 7 (SS7) and how it controls and interacts with public telephone networks and VoIP systems. SS7 is the standard communication system that is used to control public telephone networks. In addition to voice control, SS7 technology now offers advanced intelligent network features and it has recently been updated to include broadband control capabilities.

Introduction to SIP IP Telephony Systems

ISBN: 0-9728053-8-9 Price: $14.99
Author: Lawrence Harte #Pages: 117 Copyright Year: 2004
This book explains why people and companies are using SIP equipment and software to efficiently upgrade existing telephone systems, develop their own advanced communications services, and to more easily integrate telephone network with company information systems. This book provides descriptions of the function parts of SIP systems along with the fundamentals of how SIP systems operate.

Telecom Systems

ISBN: 0-9728053-9-7 Price: $34.99
Author: Lawrence Harte #Pages: 480 Copyright Year: 2004
This book Telecom Systems shows the latest telecommunications technologies are converting traditional telephone and computer networks into cost competitive integrated digital systems with yet undiscovered applications. These systems are continuing to emerge and become more complex.Telecom Systems explains how various telecommunications systems and services work and how they are evolving to meet the needs of bandwidth

Introduction to Transmission Systems

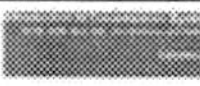

ISBN: 0-9742787-0-X Price: $14.99
Author: Lawrence Harte #Pages: 52 Copyright Year: 2004

This book explains the fundamentals of transmission lines and how radio waves, electrical circuits, and optical signals transfer information through a communication medium or channel on carrier signals. It also explains the ways that a single line can be divided into multiple channels and how signals are carried over transmission lines in analog or digital form.

Tehrani's IP Telephony Dictionary

ISBN: 0-9742787-1-8 Price: $39.99
Author: Althos #Pages: 628 Copyright Year: 2003
Tehrani's IP Telephony Dictionary, The Leading VoIP and Internet Telephony Resource provides over 10,000 of the latest IP Telephony terms and more than 400 illustrations to define and explain latest voice over data network (VoIP) technologies and services. It provides the references needed to communicate with others in the communication industry.

Practical Patent Strategies Used by Successful Companies

ISBN: 0-9746943-3-9 Price: $14.99
Author: Eric Stasik #Pages: Copyright Year: 2004

This book explains how companies can use patent strategies to achieve their business goals. Patent strategies may be considered abstract legal or economic concepts. Examining how patents are used by leading companies in specific business applications can provide great insight to their practical use and application in your business plan. This book presents in plain and clear language why having a patent strategy is important.

Introduction to xHTML

ISBN: 0-9328130-0-4 Price: $34.99
Author: Lawrence Harte #Pages: Copyright Year: 2004

This book explains what is xHTML Basic, when to use it, and why it is important to learn. You will discover how the xHTML Basic language was developed and the types of applications that benefit from xHTML Basic programs. The basic programming structure of xHTML Basic is described along with the basic commands including links, images, and special symbols.

Introduction to SS7

ISBN: 1-9328130-2-0 Price: $14.99
Author: Lawrence Harte #Pages: Copyright Year: 2004

This introductory book explains the basic operation of the signaling system 7 (SS7). SS7 is the standard communication system that is used to control public telephone networks. This book will help the reader gain an understanding of SS7 technology, network equipment, and overall operation. It covers the reasons why SS7 exists and is necessary.

Creating RFPs for IP Telephony Communications Systems

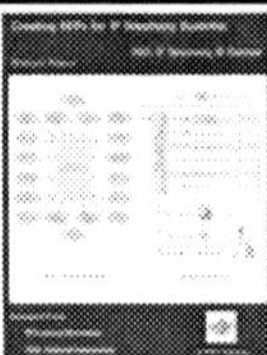

ISBN: 1-9328131-1-X Price: $19.99
Author: Lawrence Harte #Pages: Copyright Year: 2004

This book explains the typical objectives and processes that are involved in the creation and response to request for proposals (RFPs) for IP Telephony systems and services. It covers the key objectives for the RFP process, whose involved in the creation and management of the RFP, and how vendors are invited, evaluated, and notified of the RFP vendor selection result. You will learn what are RFPs and RFQs and why and when companies use and RFPs for IP Telephony Systems.

ATM Basics

ISBN: 1-9328131-3-6 Price: $29.99
Author: Lawrence Harte #Pages: Copyright Year: 2004

Asynchronous Transfer Mode (ATM) is a high-speed packet switching network technology industry standard. ATM networks have been deployed because they offer the ability to transport voice, data, and video signals over a single system. The flexibility that ATM offers incorporates both circuit and packet switching techniques into one technology.

wireless Markup Language (WML)

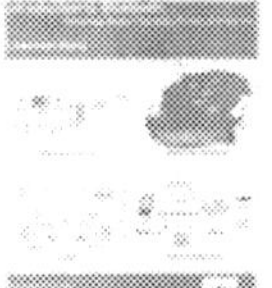

ISBN: 0-9742787-5-0 Price: $34.99
Author: Bill Routt #Pages: 292 Copyright Year: 2004

Wireless Markup Language (WML) Scripting, Scripting and Programming using WML, cHTML, and xHTML explains the necessary programming that allows web pages and other Internet information to display and be controlled by mobile telephones and PDAs.

Introduction to Bluetooth

ISBN: 0-9746943-5-5 Price: $14.99
Author: Lawrence Harte #Pages: 60 Copyright Year: 2004

Introduction to Bluetooth explains what is Bluetooth technology and why it is important for so many types of consumer electronics devices. Since it was first officially standardized in 1999, the Bluetooth market has grown to more than 35 million devices per year. You will find out how Bluetooth devices can automatically locate nearby Bluetooth devices, authenticates them, discover their capabilities, and the process used to setup connections with them.

Introduction to CDMA

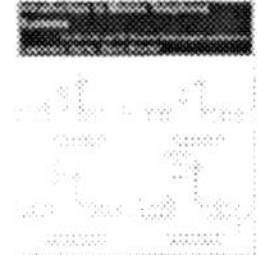

ISBN: 1-9328130-5-5 Price: $14.99
Author: Lawrence Harte #Pages: 52 Copyright Year: 2004

Introduction to CDMA book explains the basic technical components and operation of CDMA IS-95 and CDMA2000 systems and technologies. You will learn the physical radio channel structures of the CDMA systems along with the basic frame and slot structures.

Introduction to Mobile Telephone

ISBN: 0-9746943-2-0 Price: $10.99
Author: Lawrence Harte #Pages: 48 Copyright Year: 2004

Introduction to Mobile Telephone explains the different types of mobile telephone technologies and systems from 1st generation analog to 3rd generation digital broadband. It describes the basics of how they operate, the different types of wireless voice, data and information services, key commercial systems, and typical revenues/costs of these services. Mobile telephone technologies, systems, and services have dramatically changed over the past 2 years. tems to offer new services.

Introduction to Wireless Billing

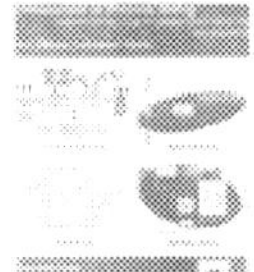

ISBN: 0-9746943-8-X Price: $14.99
Author: Avi Ofrane, Lawrence Harte #Pages: 48 Copyright Year: 2004

Introduction to Wireless Billing explains billing system operation for wireless systems, how these billing systems are a bit different than traditional billing systems, and how these systems are changing to permit billing of non-traditional products and services. This book explains how companies bill for wireless voice, data, and information services.

Order Form

ORDER FORM

Phone: 919-557-2260
800-227-9681
Fax: 919-557-2261
404 Wake Chapel Rd., Fuquay-Varina, NC 27526 USA
Email: success@Althos.com web: www.ALTHOS.com

Date:____________

Name:______________________________

Title:______________________________

Company:______________________________

Shipping Address:______________________________

City:________________ State:________ Zip:________

Billing Address:______________________________

City:________________ State:________ Zip ________

Telephone:________________ Fax:________________

Email: ______________________________

Purchase Order #____________(New accounts: please call for approval)

Payment (select): VISA ___ AMEX ___ MC ___ Check ___

Credit Card #: ________________ Expiration Date: ________

Exact Name on Card: ______________________________

Qty.	BOOK #	ISBN #	TITLE	PRICE EA	TOTAL
Book Total:					
Discounts:					
Sales Tax (North Carolina Residents please add 7% sales tax)					
Shipping: Please apply accurate shipping rates and surcharge per client and order.					
Total order:					

Providing Expert Information

Providing Expert Information

Worlds Largest Wireless Dictionary

WWW.WirelessDictionary.com

Printed in the United States
112347LV00003BB/37/A

9 780974 278780